A Field Guide to

Caribbean Reef Invertebrates

by

Nancy Sefton

and

Steven K. Webster

A Special Publication of the Monterey Bay Aquarium Foundation.

TUTA SUB AEGIDE PALLAS · EJB ·

SEA CHALLENGERS
Monterey, California

1986

E. J. BRILL
Leiden

A Sea Challengers Publication

Front cover	by Nancy Sefton
Technical Editor	William Gladfelter, Ph.D.
Copy Editor	Kenneth Hashagen

Printed by Dai Nippon Printing Co., Ltd., Tokyo, Japan

ISBN 0-930118-12-X (Sea Challengers)
ISBN 90-04-08-194-1 (E. J. Brill)

Library of Congress Catalog Card Number 85-050789

Published by SEA CHALLENGERS
4 Sommerset Rise
Monterey, California 93940

E. J. Brill
Plantijnstraat 2
Leiden

Phototypesetting and pre-press production by Padre Productions.

DEDICATION

It would be impossible for me to spend hours underwater without the support and encouragement of my husband Ron. Only his patience and willingness to help, when necessary, has allowed me to devote so much effort during the past 15 years to the photographic documentation of our local reef life. And so I dedicate this book to him, with thanks.

Nancy Sefton

Many of us can look back and attribute the setting of major courses in our lives to one or more great teachers we've had the privilege of knowing. Donald Abbott turned me (and a great many others before and since) on to the invertebrates at Stanford's Hopkins Marine Station, and a life-long passion was born. Arthur Giese had the patience and humor to hone that interest further during my graduate years at Stanford. It is to these two great teachers and valued friends that I dedicate my efforts in this book.

Steven K. Webster

TABLE OF CONTENTS

INTRODUCTION

With the increasing popularity of both snorkeling and scuba diving, more and more enthusiasts are traveling to the warm, clear waters of the Caribbean to explore nature's wondrous creation—the coral reef. This book is designed to fill a need perceived by the authors, who have spent several years teaching marine biology and providing services for divers in the Caribbean.

Once the new diver becomes physically and mentally comfortable in an alien underwater world, he or she pays increasing attention to the myriad organisms found in the coral reef community. Probably because fishes are active, colorful and familiar in other contexts, they are more readily noticed and studied than are the plants and invertebrate animals of the reef. Field guides to the fishes of the Caribbean are plentiful—some are even submersible! Thus, help abounds in identifying the common fishes. However, sport divers (and even biologists!) are awed and frequently confused by the diversity of invertebrates (and algae) they encounter. "What was THAT?" is a very common question (although difficult to understand when spoken through a snorkel). Beyond the dozens of stony corals, many of which build the framework of the reef itself, there is a rich and diverse fauna of other invertebrates and algae which live above, on, or within the reef. Still other invertebrates drift over the reef as members of the zooplankton being swept in from open water by the currents. Some, like the major reef-building corals, are large, conspicuous animals which may be a dominant feature within a particular reef habitat. Others (actually, the majority) are small, cryptic animals living in crevices, holes, and other spaces within the reef, observable only by those who know where to look.

Large or small, colorful or drab, common or rare, reef invertebrates can be a source of endless hours of enjoyable study for the diving reefwatcher and underwater photographer, especially if a compact field guide to the invertebrates is close at hand. We hope to provide such a guide between the covers of this book.

About 200 of the most common and observable of the invertebrate species (and 14 algae and 2 seedplants) are identified here. In addition, we have provided clues to the specific habitats and a few items of natural history for each species. When known, the organisms are identified to species. Because some groups are better studied than others, however, we include a few common, but unidentified, species. When scientific names vary from one author to another (some sponges, for instance, are given one name by European workers and a different one by American authors) the American choice is used out of national pride. People argue about the strangest things!

Many reef invertebrates (probably most, in fact) have been barely studied, and there is still much to learn about these organisms and their roles in the coral reef ecosystem. The authors admit to neglecting the numerous microscopic and barely visible invertebrates and algae, as these are of limited interest to most sport divers. We do, however, include a few of the more conspicuous planktonic forms frequently seen in the water column over the reef. Only a few of the most obvious and common algae and seedplants are included.

Several references are suggested for the reader who wishes to delve further into the biology and ecology of these fascinating organisms.

ACKNOWLEDGMENTS

We would like to thank Kent Schellenger whose Sierra Club trips and Sea Life Discovery programs brought us together. Thanks are also due Chuck Baxter, Judy Rand, and Dan Gotshall for their many helpful suggestions for improving the manuscript.

PHOTO CREDITS

All photographs are by the authors, except:

100 Robert G. Sellars
132, 134, 194 Dan Gotshall
154 Kent Schellenger

GLOSSARY

alga (plural=**algae**) — Photosynthetic plants that are usually composed of one cell or many similar, relatively simple cells.

amphipods — A group of small crustaceans often found living among corals and other benthic animals on the reef.

ascidian — The class of tunicates that are attached to the bottom or other substrates. May be solitary or colonial.

asexual — Reproduction or division without sex (such as coral polyps dividing to form a colony).

benthic — Living on the bottom.

buttress — Prominent seaward ends of reefs with steeply sloping, often vertical, walls.

calcareous — Composed of calcium carbonate.

carnivore — An animal which eats other animals.

cilia — Microscopic hair-like projections from a cell.

colony — A growth form whereby individuals divide to form a structure of many interconnected, usually similar, individuals. Example: colonial corals, hydroids, gorgonians, zoanthids.

commensal — A symbiotic relationship whereby one organism lives in close association with another but neither harms nor benefits its "host."

copepod — A group of extremely abundant small crustaceans. Most are planktonic and are herbivores on phytoplankton.

coralline algae — Red, green or brown algae that secrete calcium carbonate skeletal material. May help cement reef rubble together, and, after being eaten by fishes, contribute to coral sand near the reef.

cryptic — Hidden or camouflaged.

dinoflagellate — Member of a group of algae with flagella and a hard outer cell wall. Some species cause red tides. Those living in the tissues of corals and other invertebrates are the zooxanthellae.

excurrent — Flowing outward, as in excurrent openings of a sponge or tunicate.

exoskeleton — A skeleton produced outside the body of the animal which makes it (corals, arthropods, tubeworms, etc.).

flagellum A long, microscopic whiplike extension of a cell.

fouling organism Plants and animals which grow on new surfaces placed in the sea (particularly in reference to ships' hulls, pier pilings, etc.).

gorgonian Colonial members of the Class Anthozoa, often forming large fans, plumes or "bushes," each with hundreds of eight-tentacled polyps.

herbivore An animal which eats plants.

high energy reef A coral reef growing at the windward end of an island or on a shore where abundant wave action is present.

hydroids Members of the Class Hydrozoa that grow as small, often delicate polyps or colonies of polyps. Some with a free-swimming medusa in the life cycle.

hydrocorals Members of the Class Hydrozoa that secrete a calcareous exoskeleton. Includes the tropical "fire corals."

incurrent Flowing inward, as in incurrent openings of a sponge or tunicate.

jellyfish Members of the Class Scyphozoa; the true jellyfishes are free-swimming medusae, although some (the upside-down jellyfish) may rest on the bottom much of the time.

low energy reef A coral reef growing at the leeward or protected end of an island or on a protected shore with moderate to slight wave action.

lumen Central cavity or space.

medusa Free-swimming (usually), bell-like form of jellyfishes and some other coelenterates.

molt The discarded exoskeleton of an arthropod.

mysid A small (1 or 2 cm) shrimp-like crustacean often found in swarming clouds of individuals near the reef.

nematocyst A microscopic stinging structure in the tentacles and other parts of the coelenterate body.

pelagic Living in the water column; not attached to or living on the bottom.

photosynthesis The process by which algae, seedplants and other green plants capture the energy of sunlight and combine carbon dioxide and water to form complex sugars and release oxygen. This process results in the oxygen in the earth's

	atmosphere, and in the plant material that serves as food for plant-eating animals (herbivores).
plankton	An alga or animal that drifts at the mercy of the currents. Planktonic animals are called "zooplankton," while planktonic algae are termed "phytoplankton."
planula larva	The worm-like larva of a coelenterate.
polyp	The body form of a coelenterate composed of a hollow bag with a mouth surrounded by a ring of tentacles at the top. Example: anemones, corals, gorgonians, hydroids.
red tide	A massive population explosion of certain kinds of dinoflagellate algae which turns the water red or rust color.
rubble	Dead coral, shell fragments and other debris which collects on the reef.
seedplants	Flowering, seed-producing plants such as the trees, shrubs, and grasses familiar to us on land. Mangroves and turtle grass are among the few marine seedplants.
spermatozoa	Male reproductive cells (male gametes).
spicule	A skeletal structure in a sponge or coelenterate skeleton.
spur and groove system	A common feature of high energy coral reefs in which reef structures extend from the shore toward the sea in long tongues (spurs) alternating with valleys (grooves) between them.
symbiotic	"Living together." Zooxanthellae are algae living in the tissues of corals and other invertebrates and are, thus, called "symbiotic algae." Parasitism and commensalism are also examples of symbiosis.
tentacle	A long, thin extension of the body such as in coelenterates, worms and other invertebrate groups.
true corals	Members of the Class Anthozoa that grow as solitary or colonial polyps and form a calcareous (calcium carbonate) exoskeleton.
zooid	An individual member of a colony of animals.
zooxanthellae	Dinoflagellates living symbiotically with corals and other invertebrates.

CORAL REEFS OF THE CARIBBEAN

Coral reefs are geologic structures covered by shallow, tropical marine communities distributed in a band around the globe about 1500 miles above and below the equator. They are by far the largest and most extensive structures built by any organism on earth, including man. The predominant reef-builders are the "true" or "stony" corals, relatives of sea anemones, gorgonians, hydroids, and jellyfishes.

Coral reefs are usually concentrated around islands off the east coasts of continents or in other areas where warm equatorial waters rarely drop below 21° C (72° F). Reef-building corals also require shallow, sediment-free water where sunlight can support the photosynthetic activities of symbiotic algae (zooxanthellae) in the tissues of the corals. Relatively stable salinity and moderate wave action are also conducive to the formation and maintenance of coral reefs.

Geological History

True stony corals date from the Mesozoic age, about 230 million years ago. Modern Caribbean reefs, however, date back no more than 5,000 years. With the coming and going of recent ice ages, extensive changes in sea level probably left many reefs high and dry for extended periods, until sea level stabilized about 5,000 years ago.

The earliest descriptions of coral reefs originated with mariners who considered them merely dangerous obstructions to be carefully mapped and avoided. Scientific study of coral reefs began with Charles Darwin on the voyage of the *Beagle* in 1832; he described the reefs of Tahiti and other islands of the South Pacific.

Only recently has the scientific use of scuba revolutionized the study of coral reef ecology and geology. Thorough species lists, insights into the structure of reef communities and a knowledge of the biology of some of the inhabitants, now make reef systems somewhat less mysterious. Still, there is a great deal to be learned.

Caribbean Corals

About 70 species of true corals have been described for the Caribbean, compared with more than 700 species for the Indo-Pacific region. The discrepancy may be due to the younger age and less stable geology of the Atlantic and the fact that Indo-Pacific corals have been evolving longer, in a greater diversity of places, and over a much wider expanse of ocean.

The northern limit of active coral reef development in the western Atlantic is in Florida and the Bahamas, with marginal reef building as far north as Bermuda. The southern limit is on the coast of Brazil just north of Rio de Janeiro. There is an absence of reefs where the freshwater flow from the rivers forms an effective barrier to the occurrence of corals.

In the Caribbean, coral species important in the actual reef-building process include *Acropora palmata, A. cervicornis, Montastrea annularis, Diploria strigosa,* and *Porites porites*. Six coral genera and one hydrocoral (*Millepora complanata*) produce 90% of the reef framework, a situation that is fairly constant throughout the region.

Reef Structure and Zonation

Darwin described three basic types of coral reefs in the Indo-Pacific region: fringing reefs (on or near the shoreline), barrier reefs (separated from shore by a broad lagoon), and atolls (a ring of low coral islands). Caribbean reefs do not, however, fit well into this scheme, and are generally of the following types: fringing reefs, patch reefs, bank-barrier reefs, or bank (or shelf) reefs. Most Caribbean reefs are fringing or bank reefs. Bank reefs are bounded on all sides by deep water and grow on a submerged bank or platform. A bank-barrier reef grows on a submerged bank with a lagoon separating the reef from the shoreline.

Caribbean reefs may also be characterized by the amount of wave action they receive: high energy reefs grow at the windward ends of islands, and low energy reefs in less exposed areas. Whether high or low energy, reefs can be divided into specific zones distinguished by a recognizable set of dominant corals, gorgonians, and other organisms. Species composition and growth forms within these zones are determined by such factors as depth, wave action, light, geological history, and the amount of freshwater and sediment carried from nearby land areas.

A common feature of windward reefs is a spur and groove system that tends to be more uniform and well structured where there is moderate to strong wave action (at least part of the year). These spurs often terminate at their seaward ends in the spectacular buttresses, caves, tunnels, and overhangs that are the most popular, often breathtaking, sites for divers.

BANK-BARRIER REEF ZONATION. Bank-barrier reefs which have grown to near sea level usually consist of three relatively distinct zones, each with some subdivisions (if you have "splitting" tendencies). Moving shoreward from open water, the first zone is the FORE REEF or REEF FRONT. This zone extends from the seaward base of the reef to near sea level. The spur and groove system is located at the upper FORE REEF zone, with spurs and grooves oriented perpendicular to the shoreline. The most common corals of the fore reef are the elkhorn (upper zone), staghorn (lower zone), brain, and star corals. The second zone as one moves shoreward is the REEF FLAT. This shallow zone may range from 1 m to only 10 cm deep during low tide, when it may be nearly awash and impossible to snorkel over except in the occasional deeper channels through the reef. Fire corals, zoanthids, and algae dominate in this zone. Encrusting red coralline algae help to stabilize coral rubble in this area of high wave action. The inner reef flat may be somewhat deeper than the outer portion and is characterized by a considerable amount of sand and coral rubble.

The third zone is the BACK REEF, behind which is a lagoon which may be from 10 to 15 m deep and quite variable in its breadth. The lagoon generally has a sandy

A. SHALLOW FORE REEF

B. DEEP FORE REEF

13

C. REEF FLAT

D. BACK REEF

floor with small patch reefs, coral mounds or pinnacles. Green algae, turtle grass, conchs, and heart urchins are common residents.

LOW ENERGY REEFS. These reefs, growing in deeper or less exposed areas, are less well-developed than those at the more exposed windward ends of islands. Elkhorn coral, for instance, may be entirely absent from these more protected sites. Star corals, finger corals, and staghorn coral all tend to be common here, but the reef structure rarely reaches the surface. One reason for the slower reef growth typical of these leeward shores may be that lower wave energy results in more turbidity and clogging of coral surfaces with fine sediments (which also limits coral growth in lagoon and other near-shore habitats). Gorgonians are often more common on these reefs than on the windward high energy reefs.

Growth and Reproduction of Corals

Growth rates vary considerably among coral species, but are most rapid in the upright, branching Acroporas. These overtopping, treelike colonies may add 20 cm per year to the branches, while the more solid mounding and plate-like species grow considerably more slowly.

A coral polyp, thousands of which may make up a single colony, resembles a sea anemone in its structure. It is a soft, cylindrical sac crowned with a circle of tentacles around its disc-shaped top. Individual polyps vary in diameter from 1 mm to several centimeters.

A coral colony usually begins as a free-swimming or crawling planula larva released into the water by its parent polyp. This microscopic larva spends hours to days seeking an appropriate place to settle and may travel from one island to another during this time. If it survives it will settle and, with luck, find an unoccupied place to attach (such as a scar left by a grazing parrotfish). Here it will metamorphose into a minute polyp and, with its stinging tentacles, begin to feed. Almost immediately it starts to secrete a limestone exoskeleton beneath itself, thus beginning the structure of a fully developed coral colony.

When it has achieved its maximum size the polyp will divide asexually to form additional polyps just like itself; these continue to grow and divide, eventually establishing a colony of hundreds of thousands of polyps. All members of the colony are genetically identical to that first, settled polyp, and are of the same sex (well, usually. Some corals have boy and girl polyps in the same colony). In those colonies containing both sexes, self-fertilization may occur.

Most corals do not reach sexual maturity until they are about 10 years old. Spermatozoa are released into the water, often on a seasonal or monthly time schedule. They are taken in by female polyps of the same species; fertilization occurs in the central, gut-like cavity, where mature eggs are waiting. When they have developed to the planula stage, the larvae are released to the water through the mouth of the mother polyp.

Some corals undergo asexual division to form new colonies. This may occur when part of a colony breaks off during a storm, or is chewed off by a feeding fish. The fragment may settle into a crack and begin to grow anew, cementing itself to the reef in its new location. The opposite of this may occur when adjacent colonies of the same species fuse, forming a single large colony.

The longevity of corals is unknown for lack of long-term experiments. However, evidence suggests that many corals (like their sea anemone cousins) may live in excess of 100 or 200 years.

The Nutrition of Corals

True corals are of two nutritional types: those with symbiotic ("living together") algae, called zooxanthellae, and those without these microscopic algae in their tissues. Those with zooxanthellae include all of the reef-building species, as well as most non-reef-builders. The algae are now known to be crucial in the process of rapid calcium deposition and growth of the reef-builders. Without their symbionts these corals do little more than just maintain themselves. Zooxanthellae also aid in the nutrition of the corals (and in sponges, gorgonians, and anemones in which they occur). Due to the action of the zooxanthellae, reef-builders are restricted to shallow, sunlit waters where photosynthesis can take place. Their calcium deposition rates are twice on a sunny day what they are on a cloudy day.

Most of the energy on a coral reef is derived from the photosynthetic activity of zooxanthellae in corals and other coelenterates and from photosynthesis of algae in coral skeletons and algae attached to the reef itself. It is the zooxanthellae in the tissues of the corals that give them their brownish color. Only about 6 to 13% of the energy in the system enters in the form of food (plankton) eaten by reef filter-feeders and carnivores. Although coral skeletal growth rates may be several centimeters per year, overall reef growth rates result from a fine balance between constructive and destructive processes and average no more than 1 or 2 mm per year over the long haul. Reef productivity derives from four primary sources (not in order of importance): 1) filamentous algae, 2) endolithic (skeletal) algae in coral skeletons, 3) coralline algae, and 4) zooxanthellae. There are relatively few of the large algae (seaweeds) so common along the temperate shores of the world. As a result, reef herbivores have a highly productive but spatially diffuse food resource and must graze more-or-less constantly in order to make a living. Most of this productivity occurs in the sunlit waters within 10 m of the surface.

In addition to receiving nutrition from their zooxanthellae, many corals feed as carnivores, employing the stinging cells in their tentacles to capture prey. They often extend their tentacles at night when zooplankton is more abundant in the waters bathing the reef. Some corals may trap their food in sheets of mucus. The food is conducted to the mouth by microscopic cilia. Feeding on zooplankton, corals consume copepods, amphipods, worms, bryozoans, and corals. One coral colony was observed to consume 19% of its daily energy requirement in a 2-hour period after dark. For many corals, however, carnivory is not important, and most of their nutrition is provided by the zooxanthellae.

All in all, the reef should be viewed as a dynamic and delicately balanced system of great complexity and infinite variety. No two reefs are alike and years may be spent getting to "know" the reefs of just one island or coastline. We hope this book will help you recognize and appreciate the inhabitants of Caribbean coral reef communities and, in so doing, increase your skill as a naturalist and your enjoyment of diving.

INTRODUCTION TO THE MAJOR INVERTEBRATE PHYLA

PHYLUM PORIFERA—the sponges.

Although they are the least complex of all multicellular animals, sponges nevertheless have a unique set of characteristics which they share with no other groups. As the name PORIFERA (pori = hole, fera = bearing) indicates, the sponge surface is penetrated by many tiny holes leading to canals and chambers which permeate the entire sponge body. Special flagellated cells line these chambers and provide the motive power for pumping large amounts of water through the sponge. Food particles (most of them the size of bacteria and smaller) and oxygen are extracted from the water and waste products exit the sponge in the excurrent flow of water through the larger, visible holes, the oscula.

Sponges reproduce sexually as well as asexually, by fragmentation or budding. Sperm are released to the sea, sometimes in numbers so great that the sponges seem to be "smoking," and many sponges of the same species may release sperm simultaneously. While most of the sperm cells are consumed by sponges and other filter-feeders on the reef, a few enter sponges of the same species where they fertilize an egg. The resulting larva is released and swims among the plankton for some period. If it is not eaten by a filter-feeding invertebrate, the lucky larva may settle and grow in some unoccupied patch of the reef.

Some reef sponges grow several centimeters per year and are important colonizers of bare reef rock, shipwrecks, and other newly available space. Sponges, in turn, house an amazing array of commensal "guests" such as worms, shrimps, brittle stars, fishes, and algae. Some sponges bore into the limestone reef framework, weakening its structure and making it more susceptible to storm damage. Others produce extensive, nearly stony skeletal structures which cement and stabilize reef rubble and add to the structure of the reef. All combine in their nearly constant filtering activity to remove minute particles from the water and are, thus, partially responsible for the clarity of the water above the reef. A few sponges are highly toxic and should not be touched. If you aren't certain, DO NOT TOUCH! Wearing gloves will help in avoiding accidental contact with the nasty ones. Similar toxic substances probably protect many sponges from some of their potential predators, just as many land plants produce natural insecticides which protect them.

PHYLUM COELENTERATA (CNIDARIA)—corals, anemones, jellyfishes, and kin.

Although relatively simple in their basic structure, coelenterates exhibit a great variety of variations on the theme of a sac-like body with a mouth surrounded by stinging tentacles. Indeed, they are among the most widely represented of all the invertebrate phyla on the coral reef. The corals, in fact, are responsible for the reef structure itself. The phylum is divided into three classes: HYDROZOA—hydroids, fire corals, siphonophores; ANTHOZOA—corals, anemones, black corals, gorgonians; and SCYPHOZOA—true jellyfishes. All coelenterates have microscopic stinging capsules (nematocysts) with which they sting and inactivate their animal prey. A very few are dangerous to humans, but most are not.

Basic coelenterate structure is fairly straightforward. The polyp (such as an anemone or solitary coral) is a bag with a hole (mouth-anus) at the top surrounded by a ring of tentacles. Polyps are usually attached to the bottom or some other hard substrate, such as a colonial skeletal framework. Most polyps divide asexually to produce colonies (colonial corals, zoanthids, gorgonians, hydroids, etc.) consisting of hundreds to thousands of individuals. In the hydrozoa, colonies of several types of polyps may get quite complex in structure and appearance (the Portuguese Man-O-War, for instance). Colonies may be massive (corals, gorgonians) or quite small (hydroids).

Jellyfishes exhibit the MEDUSA body form, which is rather like a polyp turned upside-down with a great deal of gelatinous material in the body wall. Most medusae swim freely in the water column. Some hydroids produce small medusae. True scyphozoan jellyfishes may be large or small.

Many coelenterates contain symbiotic algae, called zooxanthellae, in their tissues. These microscopic dinoflagellates (the same group of algae which causes mussel poisoning and red tides in various parts of the world) help to nourish the coelenterate host and, in the case of corals, aid in the process of calcium carbonate secretion to form the coral exoskeleton. All reef-building corals contain zooxanthellae, as do most of the sea fans and other gorgonians. The zooxanthellae are responsible for the brownish/tan color of many of these coelenterate colonies.

Coelenterates are among the most abundant, diverse, and prominent of all reef invertebrates. Because, like the sponges, they are attached to the bottom, they are among the most easily photographed and appreciated of all invertebrate groups.

PHYLUM ANNELIDA—the segmented worms.

Most marine annelids belong to the Class Polychaeta, referring to the many bristles (chaetae, or setae) on the numerous appendages of these worms. While a few of the coral reef polychaetes are easily spotted and identified, by far the majority of the species are burrowers or nestlers and are rarely seen by the casual diver. Some may be observed at night; others by turning over pieces of coral rubble and rock (ALWAYS TURN THEM BACK THE WAY YOU FOUND THEM). Members of the group display a variety of body forms, feeding methods, and life habits. Some filter-feed while others are carnivores. Many live in tubes of their own manufacture and are permanently attached to the substrate. Polychaetes typically release sperm and eggs into the sea where fertilization results in a planktonic larva. Some species swarm together at the surface on a well-timed seasonal basis, increasing the chances that fertilization of the eggs will occur. These swarming reproductive forms may be attracted to a diver's light at night as they make their way to the surface from a hideaway in the reef below. The small, spiralling blue lights so often seen at night on the Caribbean Sea surface are examples of such spawning activities.

Because of their fast actions, polychaetes make some of the most challenging photographic subjects. We recommend great patience and some practice!

PHYLUM MOLLUSCA—the snails, bivalves, octopuses, and kin.

Of the six living classes of molluscs, three have common representatives in the coral reef community: GASTROPODA—(snails) have a single shell, or (slugs) none at all; BIVALVIA or PELECYPODA—have two shells (clams, oysters, scallops); and CEPHALOPODA—have 8 or 10 tentacular arms and a small internalized shell (octopuses and squids). This large and diverse phylum is represented on the reef by a few common and easily observed species. The majority, however, burrow in sand or nestle deep within the reef framework and will not be seen by the casual observer. Shells found on the reef or beach are best identified with one of the guides to the seashells of the Caribbean.

Sea slugs (snails without shells) of several varieties will occasionally be seen on the reef or among algal turfs in shallow, near-shore areas. Most of the shelled snails hide within the reef or burrow in sand. Only a few, like the queen conch (on sand flats) or the intertidal limpets and nerites, are commonly observed by the diver. Of the bivalves, the wing oysters and file clams are most common. Other species occur on wharf pilings, under coral rubble, and sparsely on the reef itself. Squids and octopuses are active, fast-moving predators and are behaviorally the most complex of all invertebrates. Squids, often in small schools, will be seen by day and at night, while the octopus is a solitary, nocturnal hunter and is seen only rarely during the day.

PHYLUM BRYOZOA—moss animals.

The bryozoans are small colonial animals which grow in encrusting, as well as upright, branching forms. The latter may be mistaken for hydroid colonies. Many tiny individuals, each with a crown of ciliated tentacles, make up the colony. The tentacle crown filters microscopic particles out of the water for food. Both branching and encrusting forms prefer cryptic habitats such as caves, crevices, undersides of ledges, and the dead bases of coral colonies. They are common fouling organisms on new, unoccupied substrates (like shipwrecks and toppled corals).

PHYLUM ARTHROPODA, SUPERCLASS CRUSTACEA—shrimps, crabs, lobsters, and kin.

There are more kinds of arthropods on earth than all other kinds of animals put together and the crustaceans are the most successful of the aquatic arthropods (the insects being most successful on land). Several species of mysid shrimps, true shrimps, crabs, and other groups of crustaceans are observed with ease by the diver who knows where to look. Some of the shrimps are among the most colorful and intriguing of photographic subjects and are behaviorally fascinating as well. Lobsters are becoming scarce in many areas and should be left where they are to replenish their numbers and for other divers to enjoy. All arthropods have a jointed exoskeleton which must be molted in order that the animal may grow; an occasional crab or lobster molt will be found shortly after being shed by its owner.

Vast numbers of tiny crustaceans drift in the zooplankton and are of great significance in marine food chains. They go largely unnoticed by divers, however, because of their small size. A number of the crustaceans inhabiting the reef are frequently seen at night when they emerge from their daytime hideaways and become active.

PHYLUM ECHINODERMATA—sea stars, sea urchins, brittle stars, and kin.

This phylum, with any number of truly bizarre forms, is well represented on the Caribbean reefs and in other habitats as well. They provide some of the best subjects for naturalist and photographer alike, and only the long-spined urchin bears special consideration for its uncanny ability to make contact with even the most attentive diver. Urchins belong to the CLASS ECHINOIDEA and occur in regular and irregular (heart urchins) forms. Most regular urchins scrape algae from the rocks. Heart urchins burrow in sediments and eat organic particulate matter. Feather stars (CLASS CRINOIDEA) filter plankton and other particles from the water, as do the basket stars (CLASS OPHIUROIDEA)—usually on some promontory during the dark of night. Most other ophiuroids (brittle stars) are cryptic and lie well hidden beneath rocks and in coral nooks and crannies. Sponges are another of their favorite hangouts. Sea stars (CLASS ASTEROIDEA) are not particularly common in the Caribbean reefs, but one species is commonly observed in mangroves and turtle grass beds. The sea cucumbers (CLASS HOLOTHUROIDEA) are common denizens of the reef and sand flats and make a living by vacuuming up sediments to pass through the gut.

Most echinoderms release sperm and eggs to the sea where fertilization results in planktonic larvae. These eventually settle and become juveniles if they are lucky enough to avoid being eaten by plankton-feeding animals.

PHYLUM CHORDATA, CLASS UROCHORDATA (TUNICATA) —the tunicates.

Tunicates share the Phylum Chordata with the vertebrates and are our closest relatives among the invertebrate groups. Some are gelatinous members of the plankton community, while others, the ascidians, are solitary or colonial animals adorning dead corals, rocks, vertical walls, and mangrove roots. All are filter-feeders, extracting bacteria and other tiny particles from the water as it passes over their greatly expanded gill structures. Many of the colonial ascidians (attached tunicates) form colorful patches on the reef which may be mistaken for encrusting sponges. The ascidians are generally more slick and gelatinous, and, if holes are visible, they tend to be dispersed in a regular pattern over the surface of the colony. These are the incurrent and excurrent siphons of the individual zooids which make up the colony. In addition, when touched, the holes of an ascidian colony will close rapidly, while sponges move only slowly at best.

PHYLUM PORIFERA—the sponges.

1. **WHITE CRYPTIC SPONGE** ***Leucandra aspera***
Identification: This white, calcareous sponge forms small tubes or vase-like structures with the osculum, or excurrent opening, at the top. Several of these tubes, each about 6-cm tall, may be found in a cluster. *Natural history*: *L. aspera* is found in caves and under ledges in moderate to deep locations protected from active water motion. It often grows in association with bryozoans, hydroids, and coralline red algae typical of dimly lit habitats.

2. **YELLOW CALCAREOUS SPONGE** ***Leucosolenia canariensis***
Identification: This, like most calcareous sponges, is small and grows to only about 10 cm across. It consists of intertwined yellow tubes and is structurally one of the simplest of all sponges on the reef. *Natural history*: This species is common from shallow to moderate depths in dark caves, crevices, and under large ledges.

3. **ORANGE BORING SPONGE** ***Cliona delitrix***
Identification: This bright orange sponge encrusts and bores into the calcareous skeleton of coral colonies. It may appear brown under natural light at depth (where red light is filtered out by the water). The white bumps on the sponge are the polyps of the symbiotic zoanthid, *Parazoanthus parasiticus*. The large, irregularly placed oscula rise somewhat above the surface of the sponge. *Natural history*: As a result of their boring activity, this and other boring sponges can significantly weaken coral colonies, causing them to topple more readily during storms. *C. delitrix* may encrust and kill entire coral colonies—usually the brain, star, or other massive varieties rather than the branching species.

4. **YELLOW BORING SPONGE** ***Siphodictyon coralliphagum***
Identification: Most of this boring sponge is below the surface of the calcareous skeletal substrate it inhabits. From the tubes and chambers it excavates in the coral skelelton, the sponge sends cylindrical oscula extending one to several centimeters above the surface of the coral. Incurrent openings to the sponge are in the lower yellow clumps lying near the oscula. Algae often grow on the dead coral skeleton near the sponge tissue. *Natural history*: This is but one of several sponge species which bore into and weaken coral substrates on the reef. A similar (but red) species may be *Cliona lampa*.

1. WHITE CRYPTIC SPONGE

2. YELLOW CALCAREOUS SPONGE

3. ORANGE BORING SPONGE

4. YELLOW BORING SPONGE

5. **ORANGE ELEPHANT EAR SPONGE** ***Agelas clathrodes***

Identification: This is a sometimes massive sponge forming flat orange mats which grow on or extend from vertical walls or the sides of large coral heads. The surface is smooth and rubbery, with many irregular, randomly spaced oscula. *Natural history*: Most frequently found on the deep fore reef and in canyons between buttresses, this sponge prefers dim light and quiet water. It may extend from the reef, forming a shallow bowl or ear-like flap.

6. **MOOSE ANTLER SPONGE** ***Agelas confera***

Identification: This is a large brown to tan sponge which often branches widely from a narrow base, somewhat in the form of moose antlers up to 1 or 2 m across. The surface is smooth, with randomly spaced depressions containing small symbiotic zoanthids, *Parazoanthus* sp. *Natural History*: This sponge is usually found on deep fore reef slopes and at the outer edges of buttresses. Its branching habit may restrict it to deeper, quiet water.

7. **LUMPY FINGER SPONGE** ***Agelas screptrum***

Identification: *A. screptrum* usually forms horizontal branches spreading over the reef or extending from a vertical wall. The color is orange to tan and the surface is smooth. Small depressions contain symbiotic zoanthids, while the oscula are larger and easily recognized. *Natural history*: As has been found in some other zoanthid/sponge associations, the *Parazoanthus* sp. on the surface may protect *A. screptrum* from predation by fishes while using the sponge for attachment in currents just above the reef, where feeding on zooplankton is optimal.

8. **BROWN TUBE SPONGE** ***Agelas* sp.**

Identification: One of the most common of the tubular sponges, *Agelas* sp. is usually found in large clusters, a dozen or so brown tubes growing from a broad base in moderately deep to deep reef zones. The outer surface is smooth to slightly lumpy; the osculum opens widely at the outer end of the tube. *Natural history*: Common residents of the lumen of this sponge are the yellow-and-black striped sharknose goby, *Gobiosoma evelynae*. The sponge, thus, is a highly visible cleaning station to which reef fishes come to have parasites and dead tissue removed by the cleaning gobies.

5. ORANGE ELEPHANT EAR SPONGE

6. MOOSE ANTLER SPONGE

7. LUMPY FINGER SPONGE

8. BROWN TUBE SPONGE

9. **BRANCHING VASE SPONGE** ***Callyspongia vaginalis***

Identification: This sponge forms clusters of multiple gray-to-blue tubes, 5 cm across and nearly 80 cm tall. Tubes are open at the top, and the surface is fibrous and bears many irregular conical projections. The colony may grow as a pair of nearly parallel "walls" separated by a deep channel. *Natural history*: Often covered with symbiotic zoanthids, *Parazoanthus parasiticus*, this sponge is, nevertheless, common in the diet of the gray angelfish. Brittle stars often are found draping their rays over the rims of the oscula or residing on the outer surface of the sponge.

10. **AZURE VASE SPONGE** ***Callyspongia plicifera***

Identification: This electric pink vase sponge has deep convolutions on the exterior and may be iridescent blue on the ridges and around the osculum. Growing singly or as multiple, vase-like structures, it may reach 40 cm or more in height. *Natural history*: Found from shallow to moderate depths on the reef, *C. plicifera* may adorn vertical walls in large clusters. Brittle stars and the crinoid, *Nemaster rubiginosa*, are often found in the lumen of this sponge.

11. **GIANT ORANGE SPONGE** ***Didiscus* sp.**

Identification: This is an often massive, yellow-orange sponge with a pattern of meandering grooves on the surface. These grooves contain the ostia, while the large oscula are somewhat raised and sparsely distributed over the surface. *Natural history*: Often with algae and other organisms encrusting the surface, this sponge grows most commonly on vertical walls in moderate to deep locations, typically in caves and on the walls of buttresses.

12. **BROWN VOLCANO SPONGE** ***Hemectyon ferox***

Identification: This is a sometimes massive, irregularly shaped sponge with oscula at the tops of many volcano-like cones. The surface is dull brown and felty in texture. *Natural history*: This sponge is commonly found on coral rubble and dead horizontal reef patches in moderately deep to deep locations. DO NOT TOUCH! This toxic sponge may cause considerable irritation to the skin. Indeed, it is best to wear gloves when handling sponges, as many species may cause irritation.

9. BRANCHING VASE SPONGE

10. AZURE VASE SPONGE

11. GIANT ORANGE SPONGE

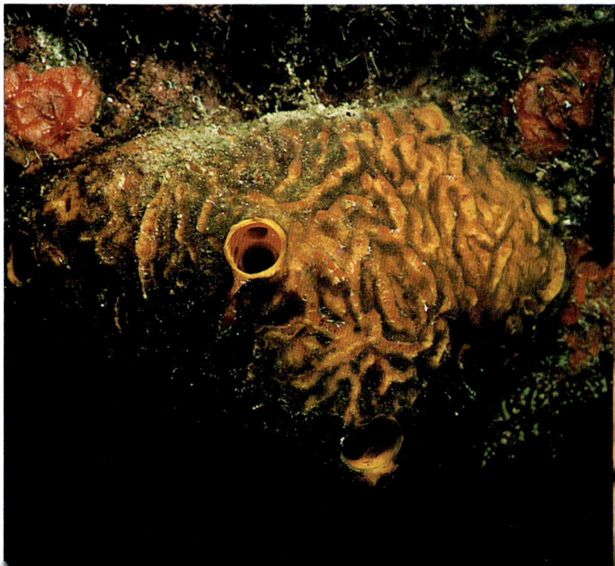

12. BROWN VOLCANO SPONGE

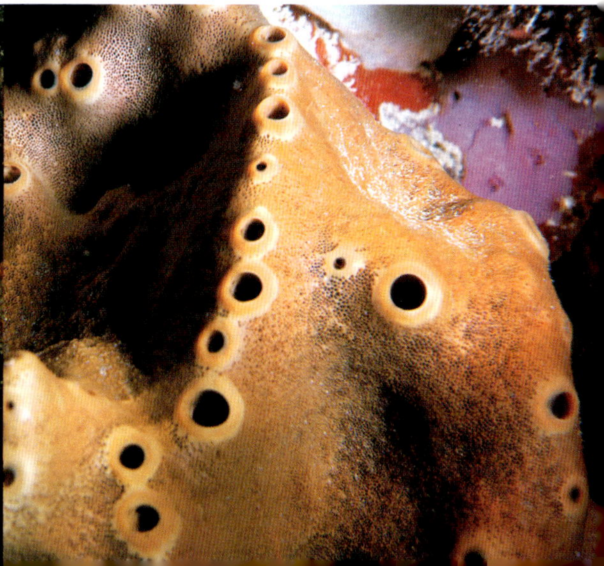

13. **ORANGE ENCRUSTING SPONGE** ***Mycale laevis***

Identification: This bright orange sponge is commonly found beneath the margins of coral plates and ledges and surrounding the base of some coral colonies. A semi-transparent membrane surrounds the large oscula which rise volcano fashion above the surface of the sponge. *Natural history*: By coating the undersides of the coral colonies on which it grows, *M. laevis* apparently protects the corals from other boring invertebrates that might otherwise settle and drill into the coral skeleton. This sponge is often found growing on *Porites* species, as well as on *Montastrea cavernosa* and *M. annularis* from shallow to deep locations.

14. **RED VASE SPONGE** ***Mycale* sp.**

Identification: Growing as single or multiple vase-like structures, *Mycale* sp. appears dark purple, or black, at depth (due to the absence of red light). At the surface, or illuminated with a photo strobe, it is scarlet red. The thick-walled tubes have a rough but gelatinous and irregular surface; it may reach 50 cm in height. *Natural history*: This sponge is found at mid-depths and deep locations on fore reef slopes, vertical walls, and on rubble in sand channels. It sometimes grows on the dead skeletons of gorgonian colonies (which may also be encrusted with fire coral). It prefers quiet water and often harbors brittle stars, which migrate to the outside at night.

15. **ORANGE LUMPY SPONGE** ***Ulosa hispida* (or *U. ruetzleri*)**

Identification: This rough-textured orange sponge encrusts dead coral and other hard surfaces on the reef. It lacks the large, prominent oscula of *Mycale laevis*, with which it may be confused. *Natural history*: Found from shallow to moderate depths on the reef, *U. hispida* is also common among the roots of the red mangrove, *Rhizophora mangle*.

16. **SMOOTH RED FINGER SPONGE** ***Haliclona rubens* (or *Amphimedon compressa*)**

Identification: The long slender branches of this sponge are usually red to orange, but may appear dull brown or black under natural light at depth. Up to 40 cm long, the branches bear small, randomly placed oscula. *Natural history*: *H. rubens* grows on shallow to deep reefs and branches upward from the reef, or outward from vertical walls. It may intertwine with other species of finger sponge. At night, and in dimly lit water, brittle stars may be seen on the surface of the sponge. Touching this sponge causes the red pigment to rub off and stain your skin or gloves.

13. ORANGE ENCRUSTING SPONGE

14. RED VASE SPONGE

15. ORANGE LUMPY SPONGE

16. SMOOTH RED FINGER SPONGE

17. **LAVENDER FINGER SPONGE** ***Haliclona hogarthi***

Identification: The slender lavender fingers of this sponge may reach a meter or more in length and may be single or multiple (often intertwined) branches. *H. hogarthi* sometimes forms large thickets 2 m, or so, across. *Natural history*: This sponge is usually found on reefs at moderate depths, but may also occur in mangrove swamps. It usually reaches its greatest size on fore reef slopes and on buttresses below the level where strong wave action is likely to occur.

18. **PINK VASE SPONGE** ***Dasychalina cyathina***

Identification: This common sponge grows to a height of 30 cm, appearing to be lavender at depth and pink at the surface. A nearly transparent membrane borders the large osculum at the top of the "vase." *Natural history*: Found at moderate depths, this sponge is often covered with the regularly-spaced polyps of the zoanthid, *Parazoanthus parasiticus*. Brittle stars may be found in the lumen, often with the tips of their rays draped over the rim of the osculum.

19. **GIANT BARREL SPONGE** ***Xestospongia muta***

Identification: Clearly the most massive of the Caribbean sponges, *X. muta* is barrel-shaped and may reach a height of 1.5 m. The large osculum may accommodate a diver (or two, or three). The surface is rough, stone hard, and may be gray or dark brown. *Natural history*: This is a slow-growing sponge which occurs in moderately deep to deep locations. The interior surface may be dotted with cleaning gobies and other small fishes which seek refuge here when threatened. The largest of these sponges are in deep dropoff areas of the fore reef slope.

20. **BOWL SPONGE** ***Cribrochalina vasculum***

Identification: This stiff, bowl-shaped sponge may be 30-50 cm across and is usually a rich, dark brown color. It feels like coarse sandpaper and sometimes fails to form a complete bowl (one side may remain open). *Natural history*: This sponge grows on the reef and on vertical walls at moderate depths. Many polyps of a small white zoanthid may dot the surface of the sponge.

17. LAVENDER FINGER SPONGE

18. PINK VASE SPONGE

19. GIANT BARREL SPONGE

20. BOWL SPONGE

21. **IRRITATING SPONGE** ***Neofibularia nolitangere***
Identification: This dark reddish-brown sponge may reach 1 m across and 30 cm or more in height. The irregular, lumpy surface is dull and felty in appearance and may be stiff (almost stony) to the touch. The large irregular openings are "cloacal openings," each containing several oscula. *Natural history*: Usually on horizontal reef or rubble surfaces, this sponge is found from shallow shore zones to the deep fore reef slope, 30 m or deeper. It harbors many commensal and parasitic animals including incredible numbers of the polychaete worm, *Syllis spongicola*, in the cloacal chambers. Vermetid gastropods (worm snails) may be seen as 1 cm diameter tubes opening at the sponge's surface. TOUCHING THIS SPONGE RESULTS IN A HIGHLY IRRITATING RASH which may last for a week or longer. DO NOT TOUCH.

22. **GREEN FINGER SPONGE** ***Iotrochota birotulata***
Identification: This is a branching sponge of irregular shape and mottled dark-to-light green color. Blotches of yellow are formed by the symbiotic zoanthid, *Parazoanthus swiftii*, which nearly always accompanies this sponge (as well as several other sponge species). The zoanthid may provide some protection against predation by fishes. *Natural history*: This sponge occurs on reefs from shallow to moderate depths, usually attached to large coral heads or vertical walls. Even with the zoanthids and their stinging abilities, *I. birotulata* is eaten by a number of fishes including the rock beauty.

23. **BLACK BALL SPONGE** ***Ircinia strobilina***
Identification: Forming a fairly massive, nearly spherical mound or ball, this sponge may grow to 0.5 m or more across. Large oscula are grouped together at the top, usually in a shallow depression. Evenly spaced, low, conical projections, with radiating white lines, cover the surface. The texture of the sponge is tough and rubbery. *Natural history*: The black ball sponge grows on and between coral heads at all diving depths, but is observed more frequently in shallow to moderately deep areas of abundant sunlight. It has been observed being eaten by the whitespotted filefish.

24. **YELLOW TUBE SPONGE** ***Aplysina ("Verongia") fistularis***
Identification: This bright yellow (or yellow-green) tube sponge retains its bright color even at depth, probably the result of fluorescent pigments. Several of these rubbery tubes may grow in a cluster and reach up to 50 cm high. The rim of each tube may bear small, finger-like projections in some specimens but not in others. *Natural history*: Abundant at moderate and deep locations on the reef, *V. fistularis* grows on reef tops, as well as on steep slopes and outcrops. Small gobies and cardinal fishes may be found resting in the lumen of the sponge. The dense network of horny skeletal fibers of this sponge make an attractive ornament.

21. IRRITATING SPONGE

22. GREEN FINGER SPONGE

23. BLACK BALL SPONGE

24. YELLOW TUBE SPONGE

25. **GIANT VERONGIA** ***Aplysina ("Verongia") gigantea***

Identification: This large yellow-to-green sponge forms massive tubes or baskets with a tough, rubbery consistency. It may reach 70 cm in height and is often encrusted with algae of various kinds. *Natural history*: *V. gigantea* prefers moderately deep to deep areas in the buttress zone and other locations free of sediment. A large specimen may be decades to centuries old, as is the case for many of the very large sponges on the reef.

26. **LONG PINK TUBE SPONGE** ***Aplysina ("Verongia") archeri***

Identification: Growing as pink-purple tubes to 2 m tall, this sponge is usually cream-color on its inner surface. The clusters of two or more tubes have a rubbery feel to them but are smooth on the outer surface. *Natural history*: This sponge is found in deep, quiet water and often branches outward and upward from a vertical wall in the buttress zone. Algae may encrust the surface of the sponge.

27. **PURPLE AND YELLOW TUBE SPONGE**
Aplysina ("Verongia") longissima

Identification: The branching, rubbery tubes of this sponge are usually yellow with varying amounts of blue and purple. Small oscula terminate each of the cylindrical branches, which reach 40 cm tall and 8 cm across. The surface is covered with small, irregular bumps. *Natural history*: Cleaning gobies will often associate with this sponge, which grows at moderate depths on the reef.

28. **PURPLE TUBE SPONGE** ***Aplysina ("Verongia") lacunosa***

Identification: The shape and size of this sponge are similar to the yellow tube sponge, except for the color. The purple color is seen inside the sponge lumen, as well as on the entire outer surface. The outer surface may be quite lumpy. *Natural history*: Clusters of these purple tubes may be massive—up to 2 m across in some deep locations where currents and wave surge are minimal.

25. GIANT VERONGIA

26. LONG PINK TUBE SPONGE

27. PURPLE AND YELLOW TUBE SPONGE

28. PURPLE TUBE SPONGE

29. **BROWN VOLCANO CARPET SPONGE** ***Anthosigmella varians***
Identification: This brown sponge forms thin, felty carpets only a few millimeters thick. Oscula are located on raised, volcano-like cones above the felty surface of the sponge. The rim of each osculum is cream-colored. *Natural history*: Encrusting carpets of this sponge may cover many square meters of rock and reef rubble in shallow, near-shore locations. The sponge appears to be immune to the effects of sedimentation (which can't be tolerated by the majority of sponge species on the reef). This sponge may also encrust coral heads and vertical walls in deeper areas, perhaps having killed the corals over which it grows.

30. **RED ENCRUSTING SPONGE** ***Monanchora barbadensis***
Identification: This very common sponge is frequently seen beneath coral ledges, on vertical walls, and encrusting the sides of coral heads. The thin red encrustations bear raised oscula with raised canals radiating from them, forming irregular star patterns on the surface of the sponge. *Natural history*: Found in moderately deep to deep locations, these encrusting red sponges may represent several species. They seem to prefer the walls of caves, overhangs, and other low-light situations.

31. **YELLOW SCLEROSPONGE** ***Ceratoporella nicholsoni***
Identification: This pale yellow sponge is rock-hard, with a dull, felty appearance. Tiny oscula are difficult to see. The sponge forms low, rounded mounds in areas of dim light (caves and deep fore reef slopes). *Natural history*: Sclerosponges belong to an ancient group of sponges thought to be extinct until their discovery in Jamaica in the 1960's. The dense calcareous base is secreted by the thin layer of living sponge on the surface of the irregular mounds. In deep water this sponge may engage in reef building at depths below which coral reef-builders are found.

32. **BROWN VOLCANO SPONGE** **Unidentified**
Identification: This large (to 60-cm diameter) sponge is similar to *Hemectyon ferox*, but the oscula lack light colored margins and are raised in tightly packed groups on relatively tall projections above the general sponge surface. The entire sponge is a bright rust-brown color. *Natural history*: This sponge is found at shallow to moderate depths, often encrusting dead coral mounds.

29. BROWN VOLCANO CARPET SPONGE

30. RED ENCRUSTING SPONGE

31. YELLOW SCLEROSPONGE

32. BROWN VOLCANO SPONGE

33. **PINK ENCRUSTING SPONGE** **Unidentified**
Identification: This sponge forms thin, encrusting pink sheets with branching canals leading to the raised oscula. *Natural history*: Found, like the red encrusting sponge, in caves, tunnels, and beneath large coral heads growing on dead calcareous reef structures.

34. **GREEN BARREL SPONGE** **Unidentified**
Identification: This large, fleshy barrel or basket shaped sponge feels rubbery and may reach 1 m high. It is green to greenish-yellow in artificial light, brown in natural light. *Natural history*: Usually found at moderately deep to deep locations on the reef.

PHYLUM COELENTERATA, Class Hydrozoa—the hydroids and hydrocorals.

35. **CHRISTMAS TREE HYDROID** ***Halocordyle disticha***
Identification: Colonies of this hydroid form white, Christmas tree-shaped colonies with alternate pinnate branches. The colonies are up to 12 cm tall. *Natural history*: These colonies are common on disturbed surfaces such as dead coral, shipwrecks, and gorgonian colonies killed by fire coral. They are also found under ledges, in caves and on vertical walls at most diving depths. The polyps of this hydroid are "naked," without a protective skeletal cup surrounding them. Extended tentacles may be seen on close inspection.

36. **BRANCHING HYDROID** ***Sertularella speciosa***
Identification: These feather-shaped colonies are, at 20 cm or so, among the largest of all the hydroids on the reef. They are evenly branched in one plane, the branches alternating along the central axis. The large polyps are usually extended, and grow alternately along the tops and bottoms of the side branches. *Natural history*: This species seems to prefer disturbed surfaces where there is some water action. It is common in caves, on shipwrecks, and on dead gorgonian colonies. Filamentous red algae are often found growing on the surfaces to which the hydroids are attached. At night you may see hermit crabs climbing the hydroid colonies, perhaps to rob the polyps of their zooplankton prey.

33. PINK ENCRUSTING SPONGE

34. GREEN BARREL SPONGE

35. CHRISTMAS TREE HYDROID

36. BRANCHING HYDROID

37. **SLENDER HYDROID** ***Cnidoscyphus marginatus***
Identification: This hydroid grows as a colony composed of a single branch about 10 cm tall. Polyps in protective skeletal cups extend alternately from opposite sides of the colony. When extended, the tentacles of the polyps give the colony a fuzzy appearance. *Natural history*: This species, like many other hydroids, often grows on dead corals and gorgonians and under coral ledges. It is photographed here growing on *Mycale* sp., the red vase sponge. Filamentous algae may grow on the hydroid exoskeleton. *Cnidoscyphus* may be found on disturbed surfaces at all diving depths.

38. **FEATHER HYDROID** ***Gynangium longicauda***
Identification: These delicate feather-like colonies grow to 15 cm, usually in small clusters. Fine branches arise alternately on opposite sides of the heavier, brown, central axis. *Natural history*: This hydroid is found in the same sorts of places as several of the other species: under coral ledges, in caves, and on vertical walls. Dead, encrusted gorgonian colonies are another typical habitat, usually where the water is free of sediment and some current is to be found.

39. **SOLITARY HYDROID** **Unidentified Corymorphid**
Identification: This large, solitary hydroid polyp grows only on the tips of branches of the sea plume, *Pseudopterogorgia* sp. Within 5 cm, or so, of the hydroid, the branch of the sea plume looks dead and no polyps are to be seen. The solitary hydroid may be 2.5 cm across; the tentacles may curl at their tips. *Natural history*: Little is known of the association of this hydroid with its gorgonian host. It is seen commonly in some locations (the Cayman Islands) and not in others (the U.S. Virgin Islands). A similar association occurs in the Sea of Cortez (Gulf of California).

40. **PLATE FIRE CORAL** ***Millepora complanata***
Identification: Fire coral colonies are beige-to-mustard color with white upper margins where growth occurs. Polyps are tiny, but extended tentacles may provide the surface with a fuzzy appearance. Colonies have a generally smooth surface in contrast to the more pitted or bumpy surfaces of the true corals. *Natural history*: Plate fire coral prefers shallow locations, particularly where active wave motion is provided by wave action. It forms dense stands on reef flats and may be nearly awash at low tides. DO NOT TOUCH. CAUSES STINGING SENSATION AND WELTS produced by the stinging cells in the extended tentacles of the defensive polyps. All three species have this capability.

37. SLENDER HYDROID

38. FEATHER HYDROID

39. SOLITARY HYDROID

40. PLATE FIRE CORAL

41. **BRANCHING FIRE CORAL**

Millepora alcicornis

Identification: This species of fire coral grows in upright branching colonies, or encrusts (and kills) sea fans and other gorgonians, taking on the shape of the affected colony. Found at all diving depths, it is the same beige color and has the smooth surface of plate fire coral. *Natural history*: This coralline hydrozoan attacks gorgonian colonies from the base and grows out along the branches (which may remain alive until actually overgrown). This fire coral and *M. squarrosa*, which forms box-like colonies in shallow water (and is uncommon in most locations), may actually be ecological variants of *M. complanata*, rather than distinct species.

42. **PINK HYDROCORAL**

Stylaster rosaceus

Identification: These delicately branched, pink colonies have white tips to their tapering branches. Tiny transparent tentacles of the protective polyps may be seen extending in clusters from visible pores on the surface of the colony. *Natural history*: *S. rosaceus* prefers moderate to shallow locations with quiet water and dim light. It is found commonly under coral ledges and arches and in caves and tunnels. The tentacles probably capture small zooplankton to feed the colony from the water passing between the many branches of the fan-shaped colonies.

PHYLYM COELENTERATA, Class Anthozoa—gorgonians, corals and anemones.

43. **ENCRUSTING GORGONIAN**

Briareum asbestinum

Identification: While most gorgonians form upright, branching growth forms, *Briareum* grows either in thick encrusting sheets which look blue and bumpy when the polyps are retracted and fuzzy and gray-brown when they are extended, or in long upright "dead man's fingers" up to 50 cm long. A closely related pink encrusting form is *Erythropodium* sp. *Natural history*: Found at all diving depths, this gorgonian encrusts the bases of coral colonies and forms small to large patches on vertical walls and in cryptic locations such as holes and crevices. It also grows on shallow sunlit reefs.

44. DEEP SEA FAN

Iciligorgia schrammi

Identification: This is one of the most distinctive of all gorgonians, forming large fan-shaped colonies at the outer ends of buttresses and on the deep fore reef slope. The thick, reddish-brown branches may or may not have their polyps extended. Extended polyps are gray-brown but, like the entire colony, appear black and white at depth. *Natural history*: Like the true sea fans, these deep sea "fans" grow across the prevailing currents so that, with polyps extended, they form effective plankton traps, feeding on small zooplankton drifting by in the currents. *Iciligorgia* may be a dominant feature of the fore reef zone at moderate and deep locations. This species is related to *B. asbestinum*; both lack the stiff woody skeleton characteristic of most gorgonians.

45. PLEXAURA ***Plexaura homomalla***

Identification: This is a dark brown, bushy gorgonian with beige-to-white polyps which, when extended, give the colony a very fuzzy appearance. *Natural history*: *Plexaura* has a very soft texture and, if squeezed, releases brownish pigment into the water. It is best not to touch this gorgonian as it MAY IRRITATE THE SKIN. It prefers shallow to moderately deep locations, particularly patch reefs and spur and groove systems. This species is the source of several pharmacologically active compounds.

46. SCRATCHY SEA WHIP

Muricea muricata

Identification: Colonies of this gorgonian are brown-to-gray with polyps extending from highly raised calices (the skeletal cups beneath the polyps). Sharp, calcareous spicules in the branches, particularly in the calices, give the colony its scratchy texture. *Natural history*: This, and other *Muricea* species, are common at moderate depths, particularly in spur and groove systems of the reef. They may also be attached to coral rubble in sandy areas.

47. KNOBBY GORGONIAN

***Eunicea* sp.**

Identification: The several species of *Eunicea* may be whip-like, branching in one plane, or bushy. Raised, almost tubular calices around each polyp produce the rough texture of the surface. These usually dark brown colonies have thick (to 1.5 cm) branches and appear fuzzy when polyps are extended. *Natural history*: Occurring at shallow and moderate depths, these gorgonians have single-celled algae (zooxanthellae) in the tissues of the polyps, as do most other gorgonians, corals, and anemones of the reef community. These symbiotic algae aid in the nutrition of the host colony.

48. BUSHY SEA WHIP

***Plexaurella* sp.**

Identification: The tall, thick branches of this gorgonian are fuzzy and brown with polyps extended but smooth with inconspicuous calices when they are retracted. Bushy sea whips are common at moderate depths growing on the reef and on coral rubble in sandy areas. *Natural history*: Coordinated action among the polyps can be demonstrated by lightly pinching a branch of the colony. A wave of retraction will move in both directions from the stimulus. Trumpet fishes will sometimes hide by aligning themselves with the branches of these colonies.

49. SEA PLUMES

***Pseudopterogorgia* sp.**

Identification: Several species of this genus form branching, plume-like colonies, some of which may reach 2 m or more in height. They may be so common as to be the dominant feature of some reefs. The color may range from gray to blue to purple. Extending polyps may be white or brownish, depending on the species in question. Of the two common species, one is slimy and one is not. *Natural history*: The polyps of the sea plumes are often retracted, giving the colonies a smooth appearance. Flamingo tongue snails are common predators of sea plumes, as well as of some other gorgonian species.

50. PURPLE SEA FAN

Gorgonia ventalina

Identification: This is the common, often large (to 2 m across) sea fan of shallow to moderate depths. Blue-to-purple in color, the colony takes on a gray-brown appearance when the polyps are extended. Some colonies may be yellow. *Natural history*: Branches of the sea fan fuse in one plane to form the large flat colonies growing perpendicular to the prevailing surges and currents (to better intercept their drifting zooplankton prey). Like true corals, many gorgonians have symbiotic zooxanthellae in their polyp tissues which provide some nourishment for their hosts.

51. GREEN SEA FAN

Gorgonia flabellum

Identification: This sea fan is yellow-green, but sometimes whitish-to-purple. Some short branches may grow at right angles to the plane of the colony. *Natural history*: The green sea fan prefers shallow reef areas with strong wave action. It is common on the windward reef flats and back reef zones where fire corals are abundant. It does not usually grow to the large size attained by the purple sea fan. Both species fall prey to the flamingo tongue snail.

52. FLAT GORGONIAN

Pterogorgia citrina

Identification: The flat, tape-like branches of this colony all lie in the same plane and are generally yellow-to-green with a thin red margin. Extended polyps are white and extend only from the margins of the branches. *Natural history*: This gorgonian is found in shallow to moderately deep areas on the reef and attached to coral rubble in sand flats and channels.

53. **ORANGE GORGONIAN** ***Ellisella* sp.**
Identification: Colonies of this red-orange gorgonian consist of slender branches with white polyps arising from distinct raised calices. The polyps arise on opposite sides of the branches. *Natural history*: This gorgonian is found only in deep water, usually in caves and tunnels or on vertical walls in quiet, dimly lit areas.

54. **BRANCHING ANEMONE** ***Lebrunia danae***
Identification: This highly variable anemone displays branched brown and white "pseudotentacles" during daylight hours, which form dense, carpet-like mats up to 20 cm across. In some forms small white inflated spheres packed with stinging capsules adorn the forked pseudotentacles. *Natural history*: Long, unbranched stinging tentacles are extended at night when planktonic prey are more abundant. *Lebrunia* is found in shallow to deep locations with just the pseudotentacles (which contain symbiotic zooxanthellae) extended from a hole or crevice in the reef or coral head.

55. **KNOBBY ANEMONE** ***Heteractis lucida***
Identification: This anemone extends its thin, pinkish tentacles from cracks or holes in reefs or coral heads. It looks superficially like *Bartholomea* (see below), except that the nematocyst batteries form raised spherical knobs over the surfaces of the tentacles. *Natural history*: These anemones are often found in coral rubble and in coral heads of the fore reef at moderate depths and deep locations. A small transparent shrimp is occasionally seen living with the anemone.

56. **GIANT ANEMONE** ***Condylactis gigantea***
Identification: Large, showy greenish-yellow tentacles with blue, pink, or purple tips make this one of the most beautiful and unmistakable anemones in the Caribbean. As is the case with most anemones on the reef, the column is hidden within the reef framework and only the tentacles extend into the water column. *Natural history*: Found in a variety of habitats (reef, lagoon, mangroves), this common anemone serves as a host to a number of commensal shrimps and fishes. In particular, look for the diamond blenny, the cleaner shrimp, and the pistol shrimp.

53. ORANGE GORGONIAN

54. BRANCHING ANEMONE

55. KNOBBY ANEMONE

56. GIANT ANEMONE

57. **CORKSCREW ANEMONE** ***Bartholomea annulata***
Identification: The long (to 10 cm) transparent tentacles of this anemone extend from holes in the reef and from under pieces of coral rubble in sandy areas. The batteries of stinging nematocysts form corkscrew-like rings around the tentacles. *Natural history*: *Bartholomea* often serves as host to Pederson's cleaning shrimp and to the pistol shrimp in moderately shallow to deep locations on the reef and in sand channels between reef spurs. Fish "cleaning" activities may often be observed around these "cleaning stations" throughout the day (so long as the observer maintains a respectful distance from the action).

58. **SUN ANEMONE** ***Stoichactis helianthus***
Identification: Carpet-like disks to 15 cm across (or more) covered with short tentacles typify this common anemone. The color varies from light tan to green. *Natural history*: *Stoichactis* generally is found in very shallow water near shore, where it is able to withstand the sediments and temperature changes typical of these areas. It often grows in dense clusters. The anemone crab, *Mithrax cinctimanus*, may be found hiding beneath the margin of the anemone.

59. **CORALLIMORPHARIAN** ***Rhodactis sanctithomae***
Identification: Polyps of *Rhodactis* may be 10 cm (but usually half this) across and cluster together to carpet areas of rock or coral rubble. Forked tentacles rise above the oral disc to branch or form flat, lettuce-like structures. The color is green with patches of blue or (sometimes) pink. *Natural history*: Found at shallow and moderate depths, the margins of these flat polyps may be turned up to form a bowl-like border to the oral disc. This action may occur rapidly and is used in food capture.

60. **GREEN CORALLIMORPHARIAN** ***Ricordea florida***
Identification: *Ricordea* forms mats of many 3- to 6-cm polyps, each with numerous short, stubby tentacles scattered over the oral disc. Colors range from green to yellow to orange. *Natural history*: Patches of these polyps are carpet-like and encrust the sides of coral heads as well as surfaces of coral rubble in shallow, near-shore areas.

57. CORKSCREW ANEMONE

58. SUN ANEMONE

59. CORALLIMORPHARIAN

60. GREEN CORALLIMORPHARIAN

61. **RED BALL ANEMONE** ***Pseudocorynactis caribbeorum***
Identification: One of the most spectacular of all anthozoans, this "red ball anemone" has transparent tentacles, each with a bright red ball at the tip. The entire "anemone" is up to 10 cm across and 10-15 cm tall. *Natural history*: These polyps are solitary and relatively uncommon. They are seen only at night extending from holes in the reef and coral rock at shallow and moderate depths.

62. **CERIANTHID** ***Arachnanthus nocturnus***
Identification: This tube-dwelling "anemone" is a solitary anthozoan seen only at night. It extends its long tentacles from the margins of the oral disc and has a ring of shorter tentacles around the mouth. The color is brownish with white bands around the tentacles. *Natural history*: These burrowing anemones extend from sand channels and sand flats at night and withdraw rapidly at the slightest increase of illumination (such as a diver's light). They live in parchment tubes, which they build with a meshwork of millions of discharged capsules.

63. **GRAY ZOANTHID** ***Zoanthus sociatus***
Identification: This common zoanthid forms carpet-like mats over extensive areas of dead coral, often on the dead inner portions of thickets of staghorn coral. The coral is variable, and may range from green, to blue, to gray, with occasional tinges of pink. Uniformly colored colonies may integrate with one another. Individual polyps are about 2 cm across with a single row of short tentacles around the margin of the disc. *Natural history*: *Z. sociatus* may be found among staghorn, finger, or lettuce corals at all but very shallow diving depths. Polyps fold their tentacles inward and retract upon being touched, and can then be seen to be connected to one another at their bases. On very shallow reefs they tend to locate in the shade.

64. **WHITE ZOANTHID** ***Palythoa caribbea***
Identification: *Palythoa* forms lumpy white to tan masses on coral rock and dead coral in shallow water. The polyps, with their single row of short tentacles, may be open or closed, at which time they look like evenly spaced constricted holes in the fleshy mass of the colony. *Natural history*: This zoanthid seems to prefer areas of abundant water action and is common in shallow fringing reef and reef flat communities. It is extremely toxic and is left strictly alone by predatory fishes (and, most likely, by everything else).

61. RED BALL ANEMONE

62. CERIANTHID

63. GRAY ZOANTHID

64. WHITE ZOANTHID

65. **GOLDEN ZOANTHID** ***Parazoanthus swiftii***
Identification: This delicate gold-to-yellow zoanthid is fairly common in moderately deep to deep areas, where it grows on green or red finger sponges. Closed polyps form small irregular yellow patches on the surface of the sponge. *Natural history*: Polyps may be open during the day in deep or well-shaded areas and are nearly always expanded for feeding on plankton at night. Zoanthids may protect the sponges on which they grow from predation by fishes. The sponge provides substrate on which the zoanthid can reside on the reef, where space is at a premium.

66. **YELLOW ZOANTHID** ***Parazoanthus parasiticus***
Identification: This zoanthid is also commonly associated with sponges, where it is found scattered more-or-less evenly over the surface of the sponge host. Connected polyps form meandering patterns on some sponges. When closed, the individual polyps are only about 3 mm across. *Natural history*: Studies demonstrate that these zoanthids discourage sponge-eating fishes, raising the possibility that this is a mutualistic (rather than parasitic) relationship benefitting both sponge host and zoanthid symbiont. Look closely at any sponge that looks fuzzy—the fuzz is probably the tentacles of the *P. parasiticus*.

67. **SPIRAL WIRE CORAL** ***Stichopathes lutkeni***
Identification: Wire coral is not easily confused with anything else. These colonies form light green to dark green spirals to 6 m long and about the diameter of a pencil. The single wire-like strand usually extends from a vertical wall, beneath a ledge, or from the roof of a cave. *Natural history*: The wire corals are composed of a single row of six-tentacled polyps running the length of the colony. They are relatives of the precious "black corals" and, like their cousins, prefer deep, quiet locations with dim light. Fore reef slopes and the walls of buttresses, 20 m deep and deeper, are the common locations.

68. **BLACK CORAL** ***Antipathes pennacea***
Identification: These large, delicately branched colonies resemble deep water ferns. The small lateral branches, which bear the feeding polyps, may appear orange or white, while the heavier main branches are thicker, black-to-rust color and are used for jewelry-making. *Natural history*: These colonies bear thousands of tiny anemone-like polyps and grow in quiet, well-shaded locations on the deep fore reef. They are becoming rare in many locations due to over-collecting for the jewelry trade.

65. GOLDEN ZOANTHID

66. YELLOW ZOANTHID

67. SPIRAL WIRE CORAL

68. BLACK CORAL

69. **BLACK CORAL** ***Antipathes* sp.**

Identification: This small, delicate black coral is found under ledges and on the roofs of caves and tunnels. The polyps are nearly always extended, each with six tentacles, in a single row on each of the thin brown-to-black branches. *Natural history*: Like most other black corals, these colonies are typically found in dimly lit, quiet locations and are often accompanied by hydroids, red coralline algae, and encrusting sponges.

70. **BLACK CORAL** ***Antipathes* sp.**

Identification: This black coral forms colonies of two or more meters high, branching in several planes to produce a large bushy appearance. *Natural history*: Like other black corals, this species is found in quiet, dimly lit areas of the fore reef slope. The several species of *Antipathes* are distinguished by distinctive branching patterns.

71. **KNOBBY CORAL** ***Madracis decactis***

Identification: These corals are brown to purplish-green and are up to 15 cm in diameter. Blunt knobs grow up from the surface of the encrusting coral mass. Sponges, tubeworms, and algae commonly grow at the bases of the knobs. *Natural history*: *M. decactis* is found in shallow, protected, back reef areas, as well as at moderate and deep zones of the fore reef slopes. Although common, it does not form a major structural element in the reef community. Polyps may be extended during the daytime.

72. **BRANCHING CORAL** ***Madracis mirabilis***

Identification: These yellow to cream-colored colonies may be several meters across and are formed of thin, delicate branches packed tightly together. The more massive colonies form gently rounded mounds. *Natural history*: This coral is found at moderate depths on buttress tops, flanks, and fore reef slopes. Brittle stars and other invertebrates are often harbored between the branches. Inner parts of the branches are dead and are encrusted with algae, sponges, and other attached invertebrates. The brittle branches of the coral are easily damaged by divers and boat anchors.

69. BLACK CORAL

70. BLACK CORAL

71. KNOBBY CORAL

72. BRANCHING CORAL

73. ELKHORN CORAL ***Acropora palmata***

Identification: The fast-growing branching colonies of this coral are sometimes 4 m or more across. The flattened (or thick and cylindrical) branches are brown to yellow with white tips (due to the lack of symbiotic algae, the zooxanthellae, in these areas of new growth). *Natural history*: One of the dominant corals in the Caribbean, elkhorn coral competes by growing rapidly and by shading or over-topping its neighbors. The most important shallow water reef-builder among the Caribbean corals, it often dominates shallow fore reef zones on windward, wave-swept shores throughout the Caribbean; it is not found below a depth of about 12 m. It is sometimes toppled by storm surf, but may re-grow from its new positions—a capability which may partially explain its wide distribution; broken fragments regenerate to form new colonies.

74. STAGHORN CORAL ***Acropora cervicornis***

Identification: Another of the fast-growing reef-builders, staghorn coral forms thickets, sometimes of great size, with a lattice-work of loosely connected branching coral colonies. Colonies may be yellow, brown, or cream color with white tips (where the growth takes place). *Natural history*: Staghorn thickets are found most often seaward of the reef flat where they may adorn the tops of buttresses at moderate depths. They are also found forming patch reefs in protected lagoons and shore zones in shallow water. Lower portions of colonies are often dead, loosely connected to other colonies (if at all), and are encrusted with algae, sponges, and tunicates. Damselfishes frequently stake out their territories in staghorn, as well as elkhorn coral.

75. LETTUCE CORAL ***Agaricia agaricites***

Identification: These colonies form leafy or encrusting tan-to-greenish colonies, often with white borders and a thin, delicate appearance. The locations of the polyps are indicated by small white star-like patterns between the low ridges which meander over the surface of the coral. *Natural history*: Lettuce coral may dominate the landscape in moderate to deep fore reef areas and buttress zones, where it forms conspicuous plate-like structures. Upright, leafy colonies are more common at moderate and shallow depths, where they may form mounding structures a meter or more across.

76. DOTTED LINE CORAL ***Agaricia grahamae***

Identification: One of several Atlantic species of *Agaricia*. The white star-shaped dots which run in the grooves of *A. grahamae* are particularly conspicuous and mark the locations of the polyps. These colonies form flat, encrusting plates and are tan to brown. *Natural history*: This coral overgrows and encrusts large areas of the reef (vertical walls, in particular) in moderately deep to deep locations of the fore reef. Sponges, algae, and tunicates often encrust the undersides of the overlapping margins of the plate-like corals. These thin plates may be fragile and are subject to damage by anchors and swim fins.

73. **ELKHORN CORAL**

74. **STAGHORN CORAL**

75. **LETTUCE CORAL**

76. **DOTTED LINE CORAL**

77. **CUP CORAL** ***Helioseris cuculata***
Identification: This small, delicate colony forms cups or shallow bowls to 20 cm across. The base color is brown, with a white outer margin and green radiating stripes. *Natural history*: Cup corals are most common at moderate to deep locations on the fore reef slope and on vertical walls. They superficially resemble some *Agaricia* species, to which they are related.

78. **PITTED CORAL** ***Siderastrea radians***
Identification: Pitted corals form small spheres or crusts. The surface of the gray or tan colony is uniformly pitted in appearance (the pits contain the withdrawn polyps of the coral). *Natural history*: This coral tolerates the sediments and the salinity and temperature fluctuations typical of shallow, nearshore areas and is found to a depth of about 10 m. It also occurs in tide pools or rocky shorelines.

79. **FINGER CORAL** ***Porites porites***
Identification: This common coral may form extensive mounds of many blunt, finger-like colonies of a tan-to-greenish or lavender color. Mounds may be several meters across and 50 cm or more tall. Polyps may be extended during the day, giving the branches a fuzzy appearance. *Natural history*: Finger coral is found in shallow back reef and reef flat zones, as well as to moderate depths on the fore reef slope. The lattice-work of branches houses nestling shrimps, brittle stars, and other invertebrates, many of which are best observed at night when they emerge from their hideouts.

80. **LUMPY PORITES** ***Porites asteroides***
Identification: This encrusting *Porites* may form mounds or flat sheets up to 60 cm in diameter. The surface is dotted with low, irregular mounds. The color may range from yellow and tan to bright green. *Natural history*: As is the case with many of the widespread coral species, *P. asteroides* exhibits different growth forms under differing conditions (such as depth). It takes a spherical or mounding shape in shallow water, but becomes flat and plate-like in deeper or shaded locations. It also grows in shallow lagoons, mangrove swamps and intertidal pools. Fan worms and Christmas tree worms often associate with this coral.

77. CUP CORAL

78. PITTED CORAL

79. FINGER CORAL

80. LUMPY PORITES

81. **SHALLOW SPHERE CORAL** ***Favia fragum***
Identification: These small, nearly spherical cream-colored colonies are usually golf-ball size and less than 10 cm across and contain just a few large polyps. *Natural history*: This coral is usually found in shallow water, often on dead coral colonies and on coral rock and rubble. It is often an early colonizer of shipwrecks and other new substrates.

82. **BRAIN CORAL** ***Diploria strigosa***
Identification: This is the most widespread of the brain corals. It forms flat to hemispherical colonies that may reach in excess of 2 m across. The color is brownish-green to yellow. The valleys (which contain the polyps) branch in a fairly regular pattern. *Natural history*: In this and other brain corals, the polyps extend their tentacles at night but keep them withdrawn during the day. *D. strigosa* occurs at all scuba depths from shallow nearshore reefs to moderately deep fore reef slopes.

83. **BRAIN CORAL** ***Diploria clivosa***
Identification: This species forms mats or hemispheres to about 1 m across. The color is green to brownish-gray. The valleys are wider than the ridges (which are narrower than in *D. strigosa*) and exhibit less branching than is characteristic of other *Diploria* species. The colony surface may be slightly lumpy. *Natural history*: *D. clivosa* grows in shallow to moderately deep areas, often in quiet back reef and lagoon habitats. Where wave action is stronger, it exhibits a more plate-like growth and becomes an important structural element of the reef community in some locations.

84. **BRAIN CORAL** ***Diploria labyrinthiformis***
Identification: One of the most distinctive of the brain corals, *D. labyrinthiformis* forms generally hemispherical colonies to more than 2 m across. The bold ridge and valley systems have a very angular appearance and a depression of uneven width runs along the ridge tops. The color is yellow to brown. *Natural history*: This common coral is found from shallow to deep locations, but is most abundant at moderate depths on windward reef terraces. It is fairly high on the scale of aggression and fends off competitors by digesting them. Polyps with a grayish color give the colony a fuzzy appearance at night.

81. SHALLOW SPHERE CORAL

82. BRAIN CORAL

83. BRAIN CORAL

84. BRAIN CORAL

85. **ROSE CORAL** ***Manicina areolata***
Identification: These small, flower-like colonies may not be attached to the substrate. They are most common in grass beds where a conical base projects down into the sand. They are yellow to brown in color, with raised, undulating margins. *Natural history*: This species tolerates shallow, sediment-rich areas near shore and grows in lagoons and sea grass beds, as well as in many reef localities.

86. **GROOVED BRAIN CORAL** ***Colpophyllia natans***
Identification: This brain coral forms mounds, hemispheres, or plates 1 m or more across, usually at moderate depths. Grooves and ridges are of different colors and are wider than in the Diplorias. The ridges are gray, the valleys green. *Natural history*: This coral is a common component of the fore reef slope, where it may be a significant reef builder.

87. **STAR CORAL** ***Montastrea annularis***
Identification: The flat plates, hemispheres, and mounds formed by this common and very important reef-builder are sometimes massive. The colonies are brown, to gray, to green. Polyps are closed during the day and are on raised calices about 3 mm across. *Natural history*: This star coral often forms massive mounds that are important structural elements of buttresses and other fore reef elements at moderate depth. Colonies become more plate-like as depth increases. This is frequently the dominant reef-builder in buttresses and fore reef slopes.

88. **STAR CORAL** ***Montastrea cavernosa***
Identification: These colonies are flat to mounding and may be 2 m or more across. Polyps are nearly dime-size and are generally retracted during the day. The color is brown to gray-green, or even bright green. Some colonies may have a rose pink tinge to them. *Natural history*: *M. cavernosa* is somewhat less common than *M. annularis* but, nevertheless, is an important reef-builder in many areas. Polyps open in dim light and at night and appear like small anemones. Deep colonies which look pink are fluorescing (taking in blue light and emitting it in the red part of the spectrum). Colonies in the deep fore reef zone are flat and plate-like, perhaps as a means of gathering as much light as possible where light is dim, at best.

85. **ROSE CORAL**

86. **GROOVED BRAIN CORAL**

87. **STAR CORAL**

88. **STAR CORAL**

89. **TAN BRAIN CORAL** ***Meandrina meandrites***
Identification: *Meandrina* grows as tan-to-white plates, mounds, or spheres with distinct meandering ridges. The septa which form the ridges are very large and easily seen by the diver. Colonies may be 1 m or more across. *Natural history*: The polyps of this coral extend their 2-cm long tentacles at night. Like a number of other corals, tan brain coral forms spheres in shallow water and flat plates in deeper locations. It has a high degree of digestive dominance (can digest and, thus, compete effectively with other coral species). It prefers shallow to deep locations on windward reefs.

90. **STARLET CORAL** ***Dichocoenia stokesi***
Identification: The polyps of this distinctive coral form white star-like patterns over the surface of this cream-color to tan colony. It may form mounds or spheres up to 50 cm in diameter. *Natural history*: This coral is found at all diving depths in back reef as well as fore reef areas. Tentacles are extended at night, giving the coral the appearance of a deep-pile carpet.

91. **PILLAR CORAL** ***Dendrogyra cylindricus***
Identification: These large, fuzzy colonies are unmistakable with their groups of vertical branches, which form spires 3 m or more tall. The white-to-tan tentacles of the polyps are extended day and night. *Natural history*: Pillar coral grows from shallow to moderate depths in relatively protected areas. It has been seen feeding on pelagic tunicates swept over the reef by currents at night. Colonies toppled by storms can re-grow vertical branches at right angles to the original axis of the colony. Their distribution is spotty throughout the Caribbean.

92. **GIANT POLYP CORAL** ***Mussa angulosa***
Identification: The polyps of this coral are among the largest of Caribbean corals. They may reach 5 cm or more across. Several polyps clustered together form mounds to about 1 m across. Polyps are green, brown, gray, or pink. *Natural history*: Short tentacles at the margin of the polyp expand only at night. This coral is digestively dominant over most other Caribbean corals and so does not become overgrown by its neighboring coral competitors. Polyps are connected by the dead basal portions of the branches on which they grow.

89. TAN BRAIN CORAL

90. STARLET CORAL

91. PILLAR CORAL

92. GIANT POLYP CORAL

93. SOLITARY CORAL

Scolymia cubensis

Identification: These corals grow only as a single coral polyp up to 10 cm or more across. *S. lacera* may grow larger and in shallower water than *S. cubensis*, which extends short tentacles around its perimeter at night. *Natural history*: Solitary corals prefer vertical walls or the undersides of ledges and larger coral colonies. They exhibit a high degree of digestive dominance over other coral species.

94. GREEN CORAL

Mycetophyllia aliciae

Identification: The flat plates of this coral form discs up to 30 cm in diameter. Low ridges radiate to the margins of the disc. The raised polyps are lighter in color than the rest of the colony. The color is gray, green, blue, or brown. *Natural history*: This attractive coral is found at moderate and deep locations of the fore reef slope. It prefers vertical walls and quiet water areas, where it is only loosely attached to the reef.

95. SCALLOPED GREEN CORAL

Mycetophyllia lamarckiana

Identification: These colonies form low mounds with scalloped margins. Gray-to-white ridges radiate outward and follow the margin of the colony, which may grow to 30 cm across. *Natural history*: This coral is found at all scuba depths, but is most common at moderate depths on outer fore reef slopes. All *Mycetophyllia* species exhibit high digestive dominance.

96. FLOWER CORAL

Eusmilia fastigiata

Identification: The branching colonies of this coral are brown, green, or yellow and have one large polyp at the tip of each branch. Colonies form mounds which may be 1 m or more across. *Natural history*: The polyps are generally retracted during the day, but extend long transparent tentacles for feeding on plankton at night. Small white bumps on the tentacles are the nematocyst batteries containing the stinging apparatus characteristic of all members of this phylum. Encrusting sponges, algae, and tubeworms often grow on the dead branches from which the polyps grow.

97. ORANGE CLUMP CORAL

Tubastrea aurea

Identification: These yellow-orange colonies have a single large polyp at the tip of each of several branches. Yellow tentacles expand at night or under low light conditions during the day. *Natural history*: *Tubastrea* grows on vertical walls, under ledges, and on shipwrecks and wharf pilings at moderately deep to deep locations. While abundant around some Caribbean islands, it is entirely absent around others.

PHYLUM COELENTERATA, Class Scyphozoa—the jellyfishes.

98. MOON JELLY ***Aurelia aurita***

Identification: This pale bluish (actually it is nearly transparent) jellyfish may reach 20 cm or more across and has a single row of numerous, fine tentacles around the margin of the bell. Four white or pink horseshoe-shaped gonads may be seen within the bell. *Natural history*: *Aurelia* is usually encountered drifting above the reef, having been swept in from the open sea. It catches small particles and plankton in its mucus covering which is transported to the mouth by cilia in special food tracks. The moon jelly is not a potent stinger and may be handled safely. Other small transparent jellies may, however, be potent stingers! Yellowtail snappers have been observed eating small *Aurelia* as they drift over the reef.

99. **UPSIDE-DOWN JELLYFISH**
Cassiopeia xamachana
Identification: This is a large (to 20 cm or more across) white-to-tan jellyfish that has the uncommon habit of lying upside-down on its bell. It prefers shallow lagoons and tidal creeks in mangrove swamps. The oral arms are highly branched and cover the entire upper (really lower) surface of the animal. It lacks marginal tentacles. *Natural history*: This true jellyfish stings small animals for food but is also nourished by the zooxanthellae (microscopic algae) that give it its tan color. Nematocyst batteries at the margin of the bell carry potent stinging cells (similar to fire coral) and should be avoided. Branched canals lead from hundreds of thousands of tiny tentacle-fringed mouthlets to the stomach.

100. **SEA WASP** ***Carybdea alata***
Identification: This most fearsome of Caribbean jellyfishes will be seen primarily at night when it swims horizontally in small "schools." It trails 4-m long pink tentacles behind it, one from each lower corner of the cuboidal bell. The transparent bell is the size of a cube of butter. *Natural history*: BEWARE! Any contact at all with the tentacles of this animal can cause severe pain and, if you're sensitive, respiratory difficulty and blackout. For this reason it is advisable to wear protective clothing which covers you completely during night dives.

PHYLUM CTENOPHORA—the comb jellies.

101. **COMB JELLY**
Unidentified Ctenophore
Identification: Comb jellies are often abundant in the plankton and look like transparent limes with eight iridescent stripes (comb rows) running from top to bottom. *Natural history*: Several species of comb jellies may be seen drifting in the currents above the reef. Morning seems to be the best time to observe these gelatinous animals as they feed on fine particles with their two hair-like tentacles (which are difficult to see without magnification). They are extremely fragile and often disintegrate when touched.

PHYLUM BRYOZOA (ECTOPROCTA) —the moss animals.

102. FAN BRYOZOAN

Unidentified Bryozoan

Identification: These small, delicate colonies are found on vertical walls, under ledges and in caves. In these protected areas they may grow to 4 cm or so across. Clusters of them look superficially like hydroid colonies and form a brownish fuzz when alive. The dead colonial skeleton turns white. *Natural history*: Bryozoans extract tiny particles from the water with their ciliated tentacles borne on numerous individual zooids. A colony is formed of hundreds of these minute individuals.

103. PINK BIRD BEAKS

Unidentified Bryozoan

Identification: This encrusting bryozoan appears light green in natural light, but is orange-pink when illuminated by strobe or diving light. Close inspection reveals a cover of beak-like projections around the individual zooids, which open to the surface of the colony. *Natural history*: These colonies form encrusting patches to 10 cm across on the walls of caves and under ledges and the sides of dead corals. Encrusting sponges and coralline algae are found in similar habitats.

PHYLUM ANNELIDA, Class Polychaeta—segmented worms.

104. GIANT FEATHER DUSTER

Sabellastarte magnifica

Identification: This largest of the feather duster worms grows most commonly from colonies of finger coral or star coral in shallow water. The "feathers" are the gills extending from the head of the worm and are brown with white bands. *Natural history*: These worms live in parchment tubes of their own manufacture and withdraw very rapidly at the least provocation. The ciliated gills not only extract oxygen from the water, but are also used in filtering fine particles which are food for these large segmented worms.

105. **SPOTTED FEATHER DUSTER** ***Branchioma nigromaculata***
Identification: The gills of *Branchioma* form a single radiating crown up to 10 cm across. They may be red or brown, banded with white. Like the giant feather duster, they tend to be solitary and grow from several kinds of coral. *Natural history*: Found in moderately deep to deep locations, these worms are extremely sensitive to disturbance and usually withdraw into their tubes one microsecond before you snap the picture (of a now-empty tube!).

106. **FEATHER DUSTER** **Unidentified Sabellid**
Identification: These small sabellids grow in small clusters with the sandy tubes clearly visible beneath the tentacular crown of gills. Gills are bluish-gray to lavender, with white tips. *Natural history*: This is one of the most easily photographed sabellids, as only a strong stimulus causes retraction of the gills into the tubes. Clusters of these worms may be uncommon, or they may cover "fields" of many square meters on coral rock in shallow locations. Like other sabellids, these worms live in tubes which they manufacture using secretions and inorganic particles.

107. **YELLOW FAN WORM** ***Hypsicomus elegans***
Identification: These colorful sabellids occur in coral rock; the crown of gills reaches about 6 cm across. These worms often occur in groups of several individuals. Gills are usually yellow, but may be white, purple, or red. *Natural history*: These worms seem to prefer vertical walls of dead coral, and other dimly lit areas at shallow to moderate depths.

108. **SPONGE WORM** ***Syllis spongicola***
Identification: These small white worms are routinely seen within the large oscula of the toxic sponge, *Neofibularia nolitangere*. The worms are parasites of the sponge and are found in several sponge species in the tropics. *Natural history*: These polychaetes are numerous within the sponge and are the preferred prey of some gobies.

105. SPOTTED FEATHER DUSTER

106. FEATHER DUSTER

107. YELLOW FAN WORM

108. SPONGE WORM

109. **TEREBELLID** ***Eupolymnia nebulosa***
Identification: Terebellids live in sandy tubes from which they extend long string-like tentacles for feeding. The tentacles often are seen extending from under pieces of coral rubble resting on sand. *Natural history*: Turning over the rubble (be sure to turn it back when you've finished worm-watching) will reveal the white worm in its tube of sand and coral fragments. The ciliated tentacles act as small conveyor belts and transport food particles back to the head of the worm. When disturbed, this worm will vacate its tube and construct another later on.

110. **FIRE WORM** ***Hermodice carunculata***
Identification: This free-crawling predator may be 30 cm long and bigger around than your little finger. It is red or brownish-green with red gills and white tufts of chaetae down both sides of the segmented body. *Natural history*: Fire worms feed on stony coral, soft coral, fire coral, and other attached invertebrates. They have calcareous chaetae which remain in your skin if you touch them. BEWARE! They can be very irritating. These worms are particularly abundant under pieces of coral rubble on sand and may be found by exploring that habitat. Be sure to turn the rocks back the way you found them.

111. **RED FAN WORM** ***Pomastegus stellatus***
Identification: Tentacles of these worms have yellow tips and form a red horseshoe about 3 cm across, often growing from star, lettuce, or finger coral. The tube is usually hidden by the crown of delicately branched tentacles. *Natural history*: Here, as in many of the tubeworms, the tube is lengthened as the coral grows, keeping the crown of feeding and respiratory tentacles just at the surface of the coral (and not overgrown by it).

112. **COLONIAL SERPULID** ***Filograna implexa***
Identification: This mass of intertwined calcareous tubes is formed by many small serpulid worms, each with a crown of orange gills. The white tubes are cemented together to form a loosely organized, somewhat spherical mass to 25 cm across. *Natural history*: More often than not these colonies are found in deep, quiet water attached to one or another species of black coral. The worms feed on tiny particles passing by in the nearly imperceptible current of these deep fore reef areas.

109. TEREBELLID

110. FIRE WORM

111. RED FAN WORM

112. COLONIAL SERPULID

113. **CHRISTMAS TREE WORM** ***Spirobranchus grandis***
Identification: One of the most common and conspicuous worms on the reef, this serpulid extends its double crown of tentacular gills above the coral heads from which it grows. These 4-cm tall "Christmas trees" come in a great variety of colors including red, yellow, white, and maroon, or a combination of two hues. *Natural history*: These worms are easily disturbed and are a challenging subject for photographers. When retracted the worm closes the tube with an ornate two-spined operculum (trap door) and reveals a sharp spine on the lip of the calcareous tube.

PHYLUM PLATYHELMINTHES, Class Turbellaria—the free-living flatworms.

114. **POLYCLAD FLATWORM** ***Pseudoceros pardalis***
Identification: This animal may be mistaken for a flat, ribbon-like sea slug, but it is really a large free-living flatworm. Up to 10 cm long, these brownish-blue worms have orange interior spots with smaller white spots around the margin. *Natural history*: Polyclads are the largest of the free-living flatworms. They are predators on smaller worms and other small invertebrates on the reef. They are not common.

PHYLUM ARTHROPODA, Superclass Crustacea—the shrimps, crabs, and lobsters.

115. **POSSUM SHRIMP** ***Mysidium* sp.**
Identification: Mysid shrimps, somewhat transparent and averaging about 5 mm in length, would scarcely be noticed on the reef if they did not hover in compact swarms. The three species in the Caribbean are not easily distinguished in the field. *Natural history*: These shrimps usually cluster in front of a dark coral recess, and may look like a swarm of larval fishes. In fact, they are food for some planktivorous fishes and, when threatened, may seek refuge among the long spines of the sea urchin, *Diadema antillarum*. At night the schools disperse.

116. **PARASITIC ISOPOD** ***Anilocra* sp.**
Identification: One of the reef's most conspicuous parasites, this isopod attaches itself to several fish host species (usually between the eyes or on the gill cover). Living on blood and fluids from the host, the isopod has a flat, segmented body with a tail fan at the posterior end. *Natural history*: Filamentous algae may grow from the surface of the isopod, giving it a fuzzy appearance. *Anilocra* will often be seen attached to the forehead of the black-barred soldierfish or beneath the eyes of butterflyfishes. Juvenile isopods may be attached to juvenile host fishes. It is not known to what extent the parasite damages the host.

113. CHRISTMAS TREE WORM

114. POLYCLAD FLATWORM

115. POSSUM SHRIMP

116. PARASITIC ISOPOD

117. **PEDERSON'S CLEANING SHRIMP** ***Periclimenes pedersoni***
Identification: This common cleaner is blue-purple and white on a transparent body. The long, waving antennae are white. The legs are blue, banded with white (the blues look purple under a strobe or diver's light). *Natural history*: This shrimp lives in association with any of several anemones (*Condylactis, Stoichactis, Lebrunea*, and, most commonly, *Bartholomea*). The shrimp actively waves its antennae to attract passing fishes, rarely moving far from the anemone "station." It may explore your finger if you extended it slowly. Cleaners are probably ridding their fish hosts of parasites, dead skin, and fungi. The shrimp gradually develops immunity to the stings of the anemone with which it lives.

118. **ANEMONE SHRIMP** ***Periclimenes yucatanicus***
Identification: This shrimp has distinctive white, purple and blue patches on its transparent and brownish body. The antennae are white banded with brown; the length of the shrimp is about 3 cm. *Natural history*: Most commonly with the anemone, *Condylactis*, this shrimp may also live with *Bartholomea* or *Lebrunea*. It is immune to the stings of the anemone and may share in its meals. Although not observed to be a cleaner of fishes, it may gain protection by looking like Pederson's cleaning shrimp, and waves its antennae in like fashion.

119. **RED NIGHT SHRIMP** ***Rhynchocinetes ringens***
Identification: Seen only at night, this shrimp grows to about 10 cm and is red with white-to-tan stripes and small spots. The large eyes are black, but they glow brightly orange in the diver's light at night. *Natural history*: This is a nocturnal animal, often seen emerging from colonies of finger coral or star coral at moderate to deep locations. The extremely long, slender antennae may help in the search for prey and in detecting the approach of potential enemies.

120. **BANDED CORAL SHRIMP** ***Stenopus hispidus***
Identification: Reaching a body length of 5 cm, the banded coral shrimp is distinguished by alternating red and white bands on the body and on the long pinchers. The slender antennae are white. *Natural history*: An active cleaner, *S. hispidus* may occur in small groups or in territorial pairs beneath coral ledges and overhangs, its presence indicated by the long, waving antennae. The authors observed one of them falling prey to an indigo hamlet (a small member of the grouper family).

117. PEDERSON'S CLEANING SHRIMP

118. ANEMONE SHRIMP

119. RED NIGHT SHRIMP

120. BANDED CORAL SHRIMP

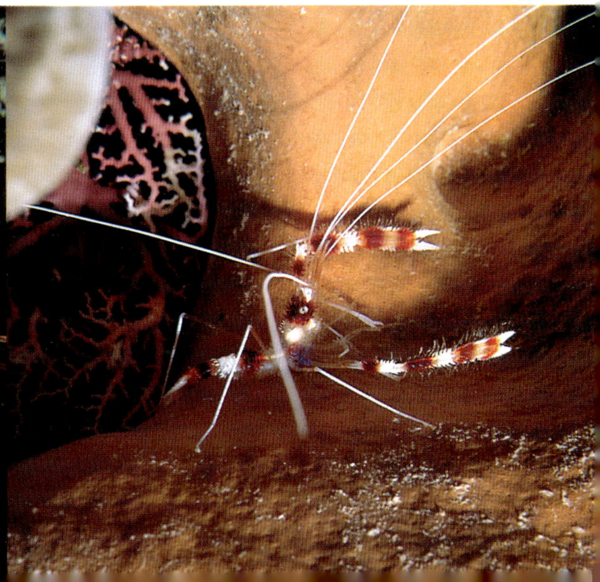

121. **YELLOW SHRIMP** ***Stenopus scutellatus***
Identification: Uncommon, this species of shrimp has alternating red and white bands on its yellow abdomen and pinchers. *Natural history*: This shrimp tends to stay hidden in small holes and under coral debris and is not seen nearly as often as the banded coral shrimp. Like its more common cousin, it is a cleaner (of parasites on fishes).

122. **SPONGE SHRIMP** ***Lysmata wurdemanni***
Identification: This shrimp, up to 4 cm long, is deep transparent pink with red longitudinal stripes. The antennae are deep pink. *Natural history*: The sponge shrimp hides inside the tubular sponges, particularly of the genus *Aplysina*, retreating deep into the lumen if disturbed. At rest just inside the osculum of the sponge, the shrimp sways from side to side with only its antennae emerging. On rare occasions this shrimp may be seen engaging in cleaning activity.

123. **ANEMONE SHRIMP** ***Thor amboinensis***
Identification: Reaching a length of only about 10 mm, tiny *T. amboinensis* has an upturned tail, giving it a teardrop shape. This shrimp is brown with white patches, white eyes, and short, banded antennae. *Natural history*: Most commonly found in small groups among the tentacles of *Condylactis gigantea*, this agile shrimp hides deep within the tentacular crown of the anemone when threatened. *Thor* may also be found among the tentacles of *Stoichactis helianthus*, sharing this host with other shrimp species.

124. **PISTOL SHRIMP** ***Alpheus armatus***
Identification: This shrimp is brown to orange to red, with mottled white markings. The long antennae are banded in red and white in this shrimp, which grows to about 3 cm in body length. *Natural history*: This secretive shrimp bears one large pincher that produces a loud snap when a vacuum is created upon the rapid opening of the pincher. This shrimp is typically found beneath *Bartholomea annulata*, with only its antennae visible among the tentacles of the anemone. The shrimp may dart forward to challenge an intruder. The snap is used to stun prey and warn other shrimps to "keep clear of my territory."

121. **YELLOW SHRIMP**

122. **SPONGE SHRIMP**

123. **ANEMONE SHRIMP**

124. **PISTOL SHRIMP**

125. **SPINY LOBSTER** ***Panulirus argus***

Identification: This is the largest and most sought-after of the Caribbean lobsters. The carapace of this large crustacean is covered with sharp spines, as are the antennae. The carapace and abdomen are brown, shading to beige on the sides and underneath. Distinctive dark spots and white markings make this a striking member of the reef community. *Natural history*: The spiny lobster is usually secretive during the day, often hiding in a hole, cave, or beneath a ledge with just the long antennae visible to the diver. Females may be seen carrying bright orange eggs beneath the abdomen during the spring. *P. argus* has been severely depleted by sport and commercial fishing in many populated areas. It feeds on reef invertebrates which are crushed or scavenged by the lobster.

126. **ROCK LOBSTER** ***Panulirus guttatus***

Identification: Somewhat smaller than the spiny lobster, *P. guttatus* has a darker body with many white spots, giving it a jewel-like appearance when observed with artificial light at night. Outer segments of the legs are striped in yellow and black. *Natural history*: A cryptic animal, the rock lobster lies well hidden in holes and caves during the day, often clinging upside-down to the roof of a small cave or ledge. It is best observed at night when it scavenges for food in the area of its daytime refuge.

127. **SPANISH LOBSTER** ***Scyllarides aequinoctialis***

Identification: This oddly-shaped lobster lacks the long antennae of the spiny and rock lobsters. A pair of flat, spatulate antennae adorn the head of this mottled beige and brown animal, which may reach a length of 30 cm. *Natural history*: This rarely seen animal crawls into the reef during the day and emerges to feed only at night. Two other "slipper lobster" species may also be encountered at night, but rarely.

128. **HERMIT CRAB** ***Dardanus venosus***

Identification: Like other hermit crabs, *D. venosus* occupies the empty shell of a gastropod (snail) for protection of its otherwise soft and vulnerable abdomen. This hermit crab is adorned with small bumps, each of which bears hair-like projections. The eyes are turquoise; the body and legs are red and white. Crab and shell together may be 6-10 cm across. *Natural history*: Most hermit crabs are scavengers, feeding on almost any dead organic material. *D. venosus* often carries the anemone, *Calliactis tricolor*, on its shell and derives protection from predatory octopuses by the presence of this stinging hitch-hiker.

125. SPINY LOBSTER

126. ROCK LOBSTER

127. SPANISH LOBSTER

128. HERMIT CRAB

129. **RED HERMIT CRAB** ***Paguristes cadenati***
Identification: This small hermit crab has a bright red body and legs, yellow eyestalks, and black to blue eyes. It occupies a gastropod shell which is often encrusted with red coralline algae (which give the shell its pink color). *Natural history*: These, like many of the smaller hermit crabs, are seen most commonly at night when they move out of their refuges to forage on the reef. Other small hermit crabs often climb hydroids and fire coral colonies at night in large numbers for some reason yet to be determined.

130. **GIANT HERMIT CRAB** ***Petrochirus diogenes***
Identification: Largest of the Caribbean hermit crabs, the giant hermit often occupies the shells of the queen conch. It has an enlarged right pincher and reaches a carapace length of about 8 cm. The body is dark brown or maroon, mottled with small, light-colored bumps. The eyes are blue; the antennae banded in red and white. *Natural history*: In addition to crawling freely on sand patches and lagoon areas (where queen conchs are found), the giant hermit may be seen resting among the branches of a gorgonian, the legs of the crab tightly gripping the branches. The crab withdraws quickly into its shell if disturbed, but will reappear within seconds if left alone.

131. **FLAT CRAB** ***Percnon gibbesi***
Identification: This small flat crab is frequently (but not necessarily) associated with the long-spined urchin from which it gains protection. The carapace is about 2.5 cm across; the color brown with yellow bands on the legs. A light luminescent band runs around the margin of the carapace. *Natural history*: This crab is usually found in shallow water hiding beneath the spines of the urchin. It feeds on the abundant filamentous algae which grow on the rocks and in the sand of shallow near-shore zones.

132. **KING CRAB** ***Mithrax spinosissimus***
Identification: A giant among Caribbean crabs, the king crab may weigh 8 lb. and have a carapace 17 cm across. The body and legs are rust brown mottled with beige or cream color. The large pinchers are a bluish-gray. *Natural history*: This crab hides in a crevice or hole during the day and emerges to feed on algae at night. It occurs at all diving depths and is actively hunted for food in some areas of the Caribbean. Because it is becoming depleted in many areas, it is the subject of research as a possible aquaculture species to replenish the stocks and satisfy the demand.

129. RED HERMIT CRAB

130. GIANT HERMIT CRAB

131. FLAT CRAB

132. KING CRAB

133. **ANEMONE CRAB** ***Mithrax cinctimanus***

Identification: This small crab has a flat, dark-colored body with long, hairy legs. It is found near the anemone *Stoichactis helianthus*. *Natural history*: *M. cinctimanus* is a speedy little crab which hides under the margin of the anemone when threatened. It is immune to the stinging nematocysts on the tentacles of the anemone. Distribution of the crab is like that of its host; predominantly in shallow near-shore areas.

134. **ARROW CRAB** ***Stenorhynchus seticornis***

Identification: This spindly crab looks like a large sea spider with its long, slender legs growing from the small, teardrop-shaped body which tapers forward to a long rostrum. The body is banded with brown and gold; the small pinchers are blue. *Natural history*: The arrow crab lives in a variety of habitats at all diving depths: under coral overhangs, in crevices, and on sponges, gorgonians, urchins and anemones. Filamentous algae may grow on the pointed rostrum of the crab.

135. **SWIMMING CRAB** ***Portunus sebae***

Identification: Like other swimming crabs, *Portunus* is easily recognized by its last pair of legs which are modified as paddle-like appendages. This species also has two large dark red spots on the carapace. *Natural history*: *P. sebae* is usually found in shallow, back reef areas and turtle grass beds, but may be encountered to moderate depths on the reef.

PHYLUM MOLLUSCA, Class Polyplacophora—the chitons.

136. **INTERTIDAL CHITONS** ***Chiton* sp. and others**

Identification: Chitons are molluscs with eight plates rather than a single shell. Several species live on rocky shores between the tidemarks, their plates often encrusted with algae and with a scaly "girdle" around the margin of the flattened, oval body. They may be 10 cm or more in length. *Natural history*: Chitons grip the rock firmly with a muscular, snail-like foot, and scrape algae with a zipper-like radula bearing many rows of very hard teeth. Teeth are replaced as they are worn away. These immobile-looking animals are abundant on some rocky shores.

133. ANEMONE CRAB

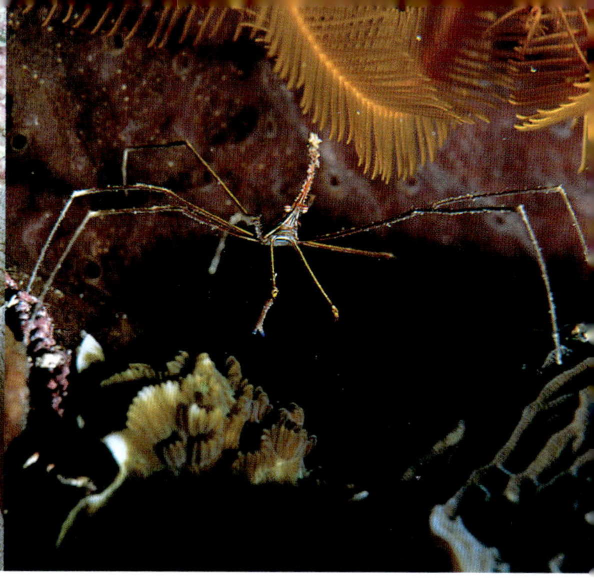

134. ARROW CRAB

135. SWIMMING CRAB

136. INTERTIDAL CHITONS

PHYLUM MOLLUSCA, Class Gastropoda—snails and slugs.

137. **QUEEN CONCH** ***Strombus gigas***
Identification: This large snail may reach a length of 30 cm. It is tan with pink or yellow inside the wide, flaring aperture of the shell. Eyestalks and the proboscis, which bears the mouth, are black and white. *Natural history*: These common conchs graze on algae and seagrasses in sand flats and shallow, sandy lagoons, particularly in turtle grass beds. Queen conchs "hop" along by extending the muscular foot and "pole vaulting" with the horny, curved operculum. The growing margin of the aperture may be thin and brittle in younger individuals. Prized for its delicate flavor, the queen conch has been severely depleted in many populated areas.

138. **TRUMPET TRITON** ***Charonia variegata***
Identification: Growing to nearly 45 cm long, the trumpet triton is one of the largest and highly prized Caribbean snails. The spiral, pointed shell is variegated beige with dark brown patches. The head bears black and yellow banded tentacles. *Natural history*: These spectacular snails are found near reefs where they may hide in holes and caves during the day. They are known to feed on sea cucumbers and are being depleted in some areas by shell collectors.

139. **MEASLED COWRY** ***Cypraea zebra***
Identification: This egg-shaped snail may be 10 cm long, with a smoothly polished, chocolate brown shell which bears white spots. The gray, fleshy mantle flaps are covered with soft, spiky projections, giving the animal a furry appearance when the flaps are extended over the shell. *Natural history*: This gastropod is occasionally seen at night crawling over a sandy bottom. Most cowries are carnivorous on encrusting invertebrates.

140. **FLAMINGO TONGUE SNAIL** ***Cyphoma gibbosum***
Identification: This small, distinctive snail reaches 2.5 cm long and is roughly oval in shape. Cream colored when the mantle flaps are withdrawn, it has bright orange spots bordered in black when the flaps are extended. The foot is yellowish with black lines. *Natural history*: This common snail is nearly always found on the sea fans or other gorgonians upon which it actively feeds. It leaves a dark trail of exposed gorgonian skeleton where it has scraped and eaten the softer tissues of its prey. Several *Cyphoma* may occur on a single gorgonian colony.

137. QUEEN CONCH

138. TRUMPET TRITON

139. MEASLED COWRY

140. FLAMINGO TONGUE SNAIL

141. **FINGERPRINT SNAIL** ***Cyphoma signatum***
Identification: Similar in size, shape, and color to *C. gibbosum*, the fingerprint snail has a mantle with parallel black stripes reminiscent of a fingerprint. *Natural history*: Somewhat rare, this snail is also found on the gorgonian colonies upon which it feeds. Two or three individuals may crowd together on a single branch of the gorgonian. The mantle flaps are withdrawn when the animal is disturbed.

142. **SACCOGLOSSAN** ***Tridachia crispata***
Identification: This snail without a shell is covered with bluish, parallel frills that are bordered with white. A pair of tentacles extends from the head of this 10 cm sea slug. *Natural history*: The saccoglossan pierces the cells of green algae and eats the contents. Algal chloroplasts are stored in the frills of the slug, where they photosynthesize and contribute to the nutrition of the "host." These animals are abundant among filamentous algae in the shallows and among algae growing on dead coral on the reef.

PHYLUM MOLLUSCA, Class Pelecypoda—mussels, clams, and oysters.

143. **PEN SHELL** ***Pinna carnea***
Identification: The thin, parchment-like shells of this bivalve are semi-transparent amber and fan-shaped, pointed at the inserted end. The shells, growing to 20 cm long, have shallow ribs running lengthwise, but this may be obscured by algae encrusting the exposed ends of the shells. *Natural history*: *P. carnea* lives almost buried in the sand or firmly nestled in a coral crevice, with only the wide ends of the shells exposed. Looking between the gaping shells one can see the gills with which the animal filters particulate matter from the water for food. The closed, encrusted shells are nearly impossible to detect.

144. **WING OYSTER** ***Pteria colymbus***
Identification: This oddly shaped oyster has an elongated hinged edge, giving it the outline of a harp. The 7-cm shell is dark, with an uneven surface and pearly interior. *Natural history*: The wing oyster attaches itself with tough byssal threads to the branch of a gorgonian colony, most often a sea plume. Here it shares the planktonic food which sweeps across the reef in the currents. Algae may encrust and obscure the shell of the oyster, which may look simply like a fuzzy bulge in the branch of the gorgonian.

141. FINGERPRINT SNAIL

142. SACCOGLOSSAN

143. PEN SHELL

144. WING OYSTER

145. **MANGROVE OYSTER**
Isognomon alatus
Identification: This flat oyster grows to 8 cm in length and occurs in clusters attached to the roots of the red mangrove. Other smaller *Isognomon* species may occur under ledges and on vertical walls on the reef. *Natural history*: Clusters of these common oysters attach to mangrove roots with byssal threads and may be covered with filamentous and other algae, as well as sponges, anemones, and tunicates. The oysters may fall prey to the predatory sea star, *Oreaster reticulatus*.

146. **ROUGH LIMA** ***Lima scabra***
Identification: This striking bivalve is distinguished by a fringe of white tentacles on the mantle margins. The white, file-like shells are up to 7.5 cm long. Internal tissues are bright red. *Natural history*: These bivalves live in coral crevices and under pieces of coral rubble on sand. Small, free-swimming individuals swim by rapid closing of the shells, taking "bites" out of the water as they move fitfully for short distances.

147. **ATLANTIC THORNY OYSTER**
Spondylus americanus
Identification: The thorny oyster has long, spiky projections growing from the margins of the shells: spines as much as half a shell-width long. Averaging 10 cm across, the shells may be encrusted with algae, hydroids, and sponges. *Natural history*: This oyster is found at moderate and deep locations on the reef; the sudden closing of the shells may be the only clue to its whereabouts. Because it is so difficult to see, the oyster is probably more common than sightings would seem to indicate.

148. **FRONS OYSTER** ***Lopha frons***

Identification: The frons oyster is easily recognized by its gaping shells, the borders of which form an angular, zig-zag pattern. The shells are to 6 cm across, and are usually encrusted with sponges, hydroids and algae. *Natural history*: These bivalves would be nearly impossible to see except for the toothy margins of the gaping shells. They grow on dead corals, as well as on gorgonian colonies and wharf pilings. When disturbed, they snap closed instantly.

PHYLUM MOLLUSCA, Class Cephalopoda—squids and octopuses.

149. **ATLANTIC OVAL SQUID**

Sepioteuthis sepioidea

Identification: This most common of Caribbean reef squids has a body length of 12 cm. The transparent tentacles form a point as the animal swims, and nearly double the apparent length of the squid. Color may mimic surrounding sand or reefs, or may turn reddish-brown or a number of other colors when disturbed. Iridescent spots occur on a dark body when disturbed at night. *Natural history*: These squids travel in pairs or small groups above the reef or sand flats. An ink "ghost" is released as the animal jets away from a threat. These squids are known to feed on fishes and small crustaceans and are themselves prey for barracuda, bar jacks, and other fast-swimming fishes.

150. **REEF OCTOPUS**

Octopus briareus

Identification: The reef octopus may reach 60 cm across from tip to tip of the arms. The body color changes with mood and surroundings, and may include blue, green, pink, rust, brown, beige, and mottled patterns. *Natural history*: A crafty hunter, the octopus moves about the reef at night probing the crevices for prey (largely crustaceans and bivalve molluscs). By day it tends to hide in a hole in the reef, the entrance of which is often bedecked with the empty shells of the octopus' prey. Males die after mating; females after brooding their eggs to the hatching stage.

PHYLUM ECHINODERMATA, Class Crinoidea—feather stars.

151. GOLDEN CRINOID

Nemaster rubiginosa

Identification: This common crinoid is easily recognized by its bright golden rays which emerge from beneath coral heads and crevices. The arms, which may reach 24 cm long, are pinnately branched and curl gracefully toward their outer ends. *Natural history*: Stiff cirri around the central disc hold the animal firmly in place on the reef. The finely branched rays bear tiny tentacular tube feet which trap plankton from the water column. When touched, the arms curl toward the disc deep in the hole or crevice in which the crinoid resides.

152. WHITE CRINOID

Nemaster discoidea

Identification: Similar to its golden cousin, the arms of this crinoid may be white, pale yellow, gray, or very dark. The food groove down the center of each arm is black. *N. discoidea* is generally smaller than *N. rubiginosa*. *Natural history*: This crinoid resides in coral crevices and under ledges at moderate and deep locations on the reef. Only a few of the 20 rays extend from the reef into the water column.

153. SWIMMING CRINOID

Analcidometra caribbea

Identification: At 10 cm across this is the smallest of the Caribbean crinoids. The 10 rays are pinnately branched and are banded in red and white. The banding may be very faint in some specimens. *Natural history*: *A. caribbea* is found in deep fore reef areas with good water circulation. It attaches to sea plumes with the cirri, which project from beneath the disc. If removed from the sea plume and released, the crinoid will swim with undulations of the rays.

154. **BLACK-AND-WHITE CRINOID**

Nemaster grandis

Identification: Each ray of this 40-armed species is up to 25 cm long. The feathery rays are densely branched and are black with white tips on the branches. *Natural history*: This crinoid, unlike its relatives, boldly exposes itself in broad daylight, often attaching to the top of a coral promontory. Usually it is found at moderate to deep locations on the reef. It may be rare or absent in some locations.

PHYLUM ECHINODERMATA, Class Ophiuroidea—brittle stars and basket stars.

155. **BRITTLE STAR**

Ophiothrix swensonii

Identification: This common brittle star has five long, thin rays radiating from the small central disc. Each ray bears a dark line down its spine, with sharp, thin transparent lateral spines radiating to either side. Colors vary from gold to brown to purple. *Natural history*: Found at all diving depths, this brittle star lives in the lumen of vase and tube sponges, as well as on some gorgonians. By day, only the tips of the rays are seen extending over the rim of the sponge. They move to more exposed positions at night, perhaps feeding on particulate matter drawn to the sponge by its filtering activity.

156. **RED BRITTLE STAR**

Unidentified

Identification: Like most brittle stars, the red brittle star is cryptic (hidden) during the day and is seen most commonly during dives at night. This large species has no spines on the rays and may be 20 cm or more from tip to tip. *Natural history*: By day this animal resides deep within the cracks and holes in the reef. Other brittle stars can be observed by turning pieces of coral rubble which rest on sand. This will also turn up file clams, terebellid worms, and fire worms (BE CAREFUL!). Be sure to return the rubble to its original position.

157. **NOCTURNAL BRITTLE STAR** **Unidentified**
Identification: This very common brittle star remains hidden by day, but is seen in great numbers extending its long, sinuous rays into the water column at night. The central disc, to which the five arms attach, remains hidden in the reef. *Natural history*: These vulnerable animals seem to avoid predation by fishes by remaining hidden during the day. They probably are plankton feeders, extracting tiny animals from the water at night. They are extremely light-sensitive, withdrawing rapidly into the reef upon exposure to a diver's light.

158. **BRITTLE STAR** **Unidentified**
Identification: This common brittle star is beige with reddish bands on the rays. Conspicuous spines project from the sides of the rays. *Natural history*: This is a common species under coral rubble and conch shells in shallow areas of the reef and sand flats. A number of other brittle star species may also inhabit these cryptic locations.

159. **BASKET STAR** ***Astrophyton muricatum***
Identification: This giant brittle star takes the tendency for branching to a ridiculous extreme. Scarcely visible during the day, it emerges at night and crawls atop a coral promontory, opening to a diameter of a meter or more. During the day it looks like a wad of straw deep within a crack or crevice in the reef. *Natural history*: The thousands of branches of the basket star form an effective plankton trap when extended across the prevailing current at night. Sensitive even to dim light, the basket star folds its rays and crawls into the reef by day or when illuminated by a diver's light at night.

PHYLUM ECHINODERMATA, Class Asteroidea—sea stars.

160. **SEA STAR** ***Oreaster reticulatus***
Identification: *Oreaster* is a large (to about 40 cm across) sea star with stout, tapering rays. Blunt spines are uniformly distributed over the upper surface. Color variations include orange, yellow, brown, red, or olive. Most individuals have five rays. *Natural history*: Probably the most commonly seen of Caribbean sea stars, *Oreaster* is found on sand flats as well as in mangroves and turtle grass beds at shallow to moderate depths. It is collected by souvenir hunters in some areas and may be locally scarce. These animals feed on bivalve molluscs.

157. NOCTURNAL BRITTLE STAR

158. BRITTLE STAR

159. BASKET STAR

160. SEA STAR

161. **COMET STAR** ***Linckia guildingii***

Identification: This distinctive sea star has finger-like arms with blunt tips and may grow to 25 cm tip to tip. It is usually brown, beige, cream, or reddish. The surface appears to be smooth, with numerous small bumps (no spines). *Natural history*: Juvenile *Linckia* may fragment, each ray then regenerating the four that are missing. The one large and four small rays give the animal its "comet" shape. Adults are rarely seen and are on sand near rocks and reefs. Young *Linckia* occur in shallow water along rocky shores.

PHYLUM ECHINODERMATA, Class Echinoidea—sea urchins.

162. **CLUB URCHIN** ***Eucidaris tribuloides***

Identification: The club urchin has relatively few, blunt, stout spines which usually are covered with sediment, filamentous algae, and dark red or pink patches of red coralline algae. The body is reddish-brown and is about 5 cm across. *Natural history*: Occurring both on near-shore flats and on the reef, this urchin remains solidly wedged in a hole or crevice by day and emerges to graze on algae at night. The encrusted spines may make the animal difficult to detect by all but the most careful observer.

163. **JEWEL URCHIN** ***Lytechinus williamsi***

Identification: The jewel-like appearance of this urchin derives from its colorful red and white body, purple-tipped pedicellariae, and green spines. The body may be 5 cm across. *Natural history*: This small urchin lives among stony corals (particularly with *Agaricia* species or *Acropora cervicornis*) on shallow reef sites. Remaining hidden in the coral thicket during the day, it may emerge somewhat to graze on algae at night.

164. **EDIBLE URCHIN** ***Tripneustes ventricosus***

Identification: This large urchin has a dark-colored body up to 15 cm across, with a dense cover of short, white spines. Bits of algae over the spines are held in place by the tube feet of the urchin. *Natural history*: This urchin may be common in turtle grass beds and in shallow water on rocky shores. The gonads (ovaries and testes) are considered a delicacy by some, as is true of several other urchin species around the world. A tiny symbiotic shrimp will sometimes be seen among the spines. The algal cover may be for camouflage or shade or it may be a food store.

161. COMET STAR

162. CLUB URCHIN

163. JEWEL URCHIN

164. EDIBLE URCHIN

165. BORING URCHIN

Echinometra lucunter

Identification: The body of this urchin is dull red or blackish, while the spines may be black, dark brown, blue, or purple. The body reaches about 8 cm in length, while the longest of the spines are about as long as the body. *Natural history*: *E. lucunter* uses its five teeth and its spines to drill into the limestone substrate creating a safe haven for itself. It emerges at night to graze algae on the shallow rocky shores and in tidepools. Urchins may defend their "home" burrows against their aggressive neighbors (other *E. lucunter*). There appears to be a "home burrow" advantage and aggressors are usually turned away.

166. REEF URCHIN

Echinometra viridis

Identification: The reef urchin has a red to maroon body. The spines are green with brownish tips and a white line around the base. Spines are shorter than the 5-cm diameter of the body, but are relatively longer than those of *E. lucunter*. *Natural history*: Reef urchins live in shallow to moderately deep locations, usually deep within coral thickets and crevices during the day. They most likely graze on algae, most actively at night.

167. LONG SPINED URCHIN

Diadema antillarum

Identification: The most common of the Caribbean urchins, *Diadema* has long (to 30 cm, or so), finely pointed spines. The body is about 10 cm across. Spines of the juvenile may be banded in black and white. An occasional adult has gray spines. *Natural history*: Hiding well within the reef by day, *Diadema* emerges to graze at night, often migrating from the reef some distance onto the sand flats. Mysid shrimps, cardinalfishes, and an occasional arrow crab may be seen among the spines of the urchin during the day. Venom glands on the surface make a spine puncture somewhat painful. Fragments left in the wound soon dissolve. Apply first aid to avoid infection.

168. RED HEART URCHIN

Meoma ventricosa

Identification: This urchin is shaped like a somewhat flattened egg with a cover of numerous thin, short spines. Up to 18 cm long, *Meoma* is usually dark red to brown in color. The five-pointed star pattern on the upper surface indicates the relationship of the sea urchins to the (usually) five-armed sea stars. *Natural history*: Living heart urchins (there are several dozen species in the Caribbean) burrow beneath the sand and are only rarely seen. Dead skeletons (tests) are more commonly seen on the surface after the urchin has fallen prey to a helmet snail (*Cassis* sp.).

PHYLUM ECHINODERMATA, Class Holothuroidea—sea cucumbers.

169. SOFT SEA CUCUMBER

Astichopus multifidus

Identification: Reaching a length of 45 cm, this sea cucumber is white with many small, black, pointed tube feet. Some specimens may be brown with white spots and blotches. The underside is flattened and bears the tube feet with which the animal moves about. *Natural history*: These very soft cucumbers live on sandy substrates which they ingest, digesting out the organic material as it passes through the gut. Cylindrical masses of clean sediment are left behind as the animal moves and feeds.

170. SEA CUCUMBER

Holothuria mexicana

Identification: This 40-cm cucumber looks segmented as a result of the heavy transverse wrinkles in the leathery body wall. It is a mottled gray, brown, black, and beige, often with a light coating of white sediment. Dark colored tube feet extend from the flattened underside for attachment and locomotion. *Natural history*: These animals will be found on or near sand, often in turtle grass beds. The mouth is surrounded by tentacular tube feet which help it extract edible organic material from the sand.

171. TIGER'S TAIL

Holothuria thomasae

Identification: This elusive sea cucumber extends its long (to 2 m or more), elastic body from the reef onto the sand at night to feed. The body is soft, about 3 or 4 cm across and covered with soft, pointed tubercles. The color is gray, brownish, or maroon with whitish blotches. *Natural history*: These animals hang on firmly to the reef with their posterior ends while sweeping the surface of the sand, vacuum-cleaner fashion, to feed on organic particles. Any slight stimulus causes them to withdraw rapidly into the reef.

PHYLUM CHORDATA, Subphylum Tunicata—the tunicates.

172. REEF TUNICATE

Ascidia nigra

Identification: This solitary tunicate appears black under natural light but is dark purple under a strobe or diver's light. The outer tunic is smooth; the incurrent and excurrent siphons are clearly visible at the top of the animal, which may reach 5 cm tall. *Natural history*: *A. nigra* nestles in a coral crevice or other cryptic habitat, but may also be seen out in the open. When disturbed, the animal stops filtering water for its particulate food, closes the siphons, and withdraws somewhat into its hideaway.

173. YELLOW TUBE TUNICATE

Ascidia sydneiensis

Identification: This large solitary tunicate is widespread throughout the tropics and is easily recognized once it is spotted. The tubular yellow incurrent siphon looks and feels gelatinous as it extends from a reef crevice, often just above a sandy bottom. It measures several cm long and about 3 or 4 cm across. *Natural history*: *Ascidia* withdraws its siphon if disturbed. Like other tunicates, it filters bacteria and tiny particles from the water for food, using its enlarged gill basket as a filter.

174. LIGHTBULB TUNICATE

Clavelina picta

Identification: This distinctive translucent tunicate looks like a small lightbulb on the reef. The incurrent siphon may be ringed with white and/or purple. Several individuals may grow from a common basal stolon network, forming a cluster of a dozen or more individuals. *Natural history*: Groups of just a few individuals may be found on vertical walls and under ledges at moderate to deep locations. Colonies of many individuals are sometimes found on mangrove roots or on dead gorgonian colonies in shallow areas.

175. BLUE ENCRUSTING TUNICATE

Unidentified Didemnid

Identification: This blue-green compound (many tiny individuals sharing a common gelatinous tunic) tunicate encrusts dead coral and gorgonian colonies. The tunicate is stiff, smooth, and gelatinous to the touch. *Natural history*: Didemnids may kill the corals and gorgonians on which they grow and may encrust large expanses of dead coral and rubble. Small spots indicate the siphons of the individual animals which comprise the colony.

176. STRAWBERRY TUNICATE

Distaplia stylifer

Identification: These small compound tunicates may be 2 to 3 cm across and are a dull red color. Each nearly spherical colony contains many small individual "zooids," or individuals. *Natural history*: These small, globular colonies grow under ledges, in crevices, and on coral rubble.

177. **PURPLE TUNICATE** ***Clavelina* sp.**
Identification: Similar in appearance to *C. picta*, these social tunicates form clusters of many attached individuals, each with a bluish-purple color in a translucent tunic. Individuals are 2 to 3 cm tall; a cluster may be 30 cm across. *Natural history*: Like *C. picta*, these colonies grow in shallow water attached to gorgonians, dead coral, and other hard substrates.

178. **GIANT TUNICATE** **Unidentified**
Identification: This large, solitary tunicate extends its siphons from a coral crevice. The tunic is beige and leathery, but is often encrusted with algae. It is so hard to spot that it may be noticed only when it closes its siphons. *Natural history*: These animals are sensitive to unusual stimuli and will close the siphons and withdraw rapidly (for a tunicate) if an intruder is detected.

179. **GREEN SOCIAL TUNICATE**
Symplegma viride
Identification: These small green individuals, each about 5 mm across, cluster together from a common stolon network and encrust small patches of dead coral on vertical walls and under ledges. *Natural history*: These colonies are found at moderately deep and deep locations where you would expect to see pink hydrocoral, hydroids, and other "under ledge" denizens. Red coralline algae and encrusting sponges are also typical of these habitats.

DIVISION CHLOROPHYTA—the green algae.

180. **GREEN PARASOL** ***Acetabularia* sp.**

Identification: *Acetabularia* is a syncytium (many cells, but with no cell walls between them) forming a 6 to 8 cm stalk with a parasol-like cap about 1 cm across. A light green color, the cap is sculptured with regular grooves which run radially from the center to the margin. *Natural history*: Clusters of these distinctive green algae grow on rock and rubble in shallow water, as well as on the shells of the queen conch in turtle grass beds and sand flats.

181. **GREEN ALGA** ***Avrainvilla* sp.**

Identification: The flat blade of this green alga is dark green. It grows to 10 cm or larger. *Natural history*: This common alga grows in shallow sand and turtle grass beds. Calcareous skeletal elements in the alga help to give it support and hold it upright out of the sediments.

182. **ARTICULATED GREEN ALGA** ***Halimeda copiosa***

Identification: One of several species of *Halimeda*, this has flat, rounded sections 2 cm or more across. The sections grow in strings hanging 50 cm or more from the reef. *Natural history*: This species of *Halimeda* grows on deep fore reef slopes from 20 m to well below scuba depths. Like other *Halimeda* species, it may contribute significant amounts of calcareous material to the reef and to sand deposits in lagoons.

183. **CACTUS ALGA** ***Halimeda opuntia***
Identification: The many segmented branches of this common *Halimeda* species grow in many directions, forming dense clumps or mats. Segments may be 2 cm across. *Natural history*: Found at all scuba depths, this alga forms mats many meters across at the bases of staghorn thickets and in other fore reef areas in moderately deep to deep locations. Filamentous red algae often grow among the branches, as do many kinds of nestling invertebrates.

184. **NEPTUNE'S SHAVING BRUSH**
Penicillus capitatus
Identification: This erect green alga has a tuft of bristle-like branches at the top and may grow to 15 cm, or more, tall. It is green, with white calcareous skeletal material on the surface. *Natural history*: *P. capitatus* grows in sand at all scuba depths, but will most commonly be seen in shallow water in turtle grass beds and near mangroves.

185. **SILVER BALL ALGA**
Valonia ventricosa
Identification: This large, single, fluid-filled cell may be 5 cm across and is green to silvery. Sediment and other algae may be found encrusting the surface of the spherical plant. *Natural history*: *Valonia* grows singly or in small clusters in coral crevices or on the dead bases of some coral colonies (such as staghorn corals). Lacking roots, this and other attached algae form a many-cell "holdfast" structure with which they anchor themselves to the substrate.

186. **GREEN FAN ALGA** ***Udotea* sp.**
Identification: Several species of *Udotea* occur on and near Caribbean reefs. The upright, flat blade is supported on a stalk which grows from the sand. The plant is stiffened by the calcareous skeletal elements secreted by the cells. *Natural history*: Skeletal elements in the calcareous algae may give mechanical support and protect the alga from grazing by some herbivorous fishes. When the alga dies (or is eaten) the calcareous skeletal material is added to the sand and sediments of the reef system.

DIVISION RHODOPHYTA—the red algae.

187. **RED ALGA** ***Daysea* sp.**
Identification: This very delicately branching red alga may look more like a hydroid colony than the alga it is. Fine red-orange tufts grow along the white branches which grow alternately from a central stalk attached to the substrate. The entire plant is no more than 8 cm long. *Natural history*: This alga grows on vertical walls, under ledges, and on dead gorgonian colonies at moderate to deep locations on the reef. It occurs with red encrusting sponges, red coralline algae, hydroids, and other typical under-ledge species.

188. **RED CORALLINE ALGA** **Unidentified**
Identification: Red coralline algae often form pink or purplish-red crusts on rocks and dead coral. They look like thin coatings of pink frosting and may cover large expanses. *Natural history*: The encrusting red coralline algae are important in cementing and stabilizing reef rubble and in consolidating other loose elements of the reef. They add appreciable amounts of calcium carbonate to the reef sediments when eaten by herbivorous fishes and other reef herbivores. In addition to growing under ledges, beneath rubble, and on vertical walls at all scuba depths, some species also form massive ridges in very shallow, near shore habitats.

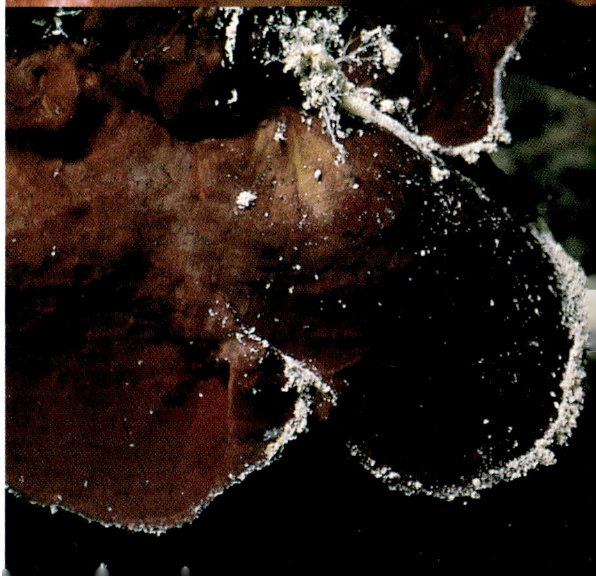

189. **RED FILAMENTOUS ALGA**
Unidentified

Identification: Filamentous red algae form reddish to brown mats and tufts on dead coral, dead gorgonians, bivalve shells, and other hard substrates (including some *Halimeda* species). *Natural history*: Perhaps among the first "fouling" organisms to grow on newly available substrate, filamentous red algae of many different species are found at all diving depths on virtually all hard, non-living substrates. Many red algae do particularly well in deep areas of low light intensity.

DIVISION PHAEOPHYTA—the brown algae.

190. **BRANCHING TAPE ALGA**
***Dictyota* sp.**

Identification: This brown alga is usually greenish, and forms mats of tape-like, branching structures in one plane. They also may exhibit an iridescent bluish color. *Natural history*: Several species of *Dictyota* grow on vertical walls, under ledges, and on dead corals at moderately deep to deep zones of the reef. Hydroids and red filamentous algae are often seen nearby or intermingled with the *Dictyota*.

191. **COMMON BROWN ALGA**
Lobophora variegata

Identification: This alga grows in upright or prostrate fan-shaped forms with concentric ridges over the entire surface. It is light brown to whitish. *Natural history*: *Lobophora* is among the most common of the shallow water algae growing in deeper locations on vertical walls and on the walls and roofs of caves.

192. **SARGASSUM WEED**

***Sargassum* sp.**

Identification: Several species of *Sargassum* occur in the Caribbean. They tend to grow attached to coral rock at shallow to moderate depths, with branches extending into the water column. Small, spherical, gas-filled floats (or bladders) keep the leaf-like blades upright in the water. The color is light-olive to brown. *Natural history*: One species, *Sargassum natans*, floats freely in the Atlantic and is probably the most abundant plant on earth. The near-shore species begin life attached with a holdfast, but portions of the plant may break loose and live for some time in the floating (pelagic) state well away from land.

193. **BROWN FAN ALGA**

***Stypopodium* sp.**

Identification: Flat, fan-shaped branches of this plant are greenish to light brown. The branches may reach more than 30 cm long, and split as they grow. Green concentric rings occur every few centimeters along the branches. *Natural history*: These are common plants in some locations at all diving depths. Shallow, near-shore rocky areas are particularly likely places to look.

DIVISION ANTHOPHYTA—the seed-plants.

194. **MANATEE GRASS**

Syringodium filiforme

Identification: Frequently observed growing with turtle grass in shallow to moderately deep sand flats, manatee grass has narrow, cylindrical leaves which may reach a length of 40 cm or more. *Natural history*: The leaves of the plant are connected by a mat of horizontal stems (stolons) just beneath the surface of the sand. Five leaves typically grow together at each node along the stolon.

195. **TURTLE GRASS**

Thalassia testudinum

Identification: The most common of the Caribbean subtidal seedplants, turtle grass grows as flat, tape-like leaves rooted in the sandy sediments. Leaves are commonly coated with fine, powdery white sediments. *Natural history*: This highly productive plant forms expansive meadows in shallow, protected back reef and lagoon areas where it is eaten by green turtles and many herbivorous fishes.

SUGGESTED REFERENCES

Atlantic Reef Corals, F. G. Walton Smith. University of Miami Press, Coral Gables, Florida. 1971.

Caribbean Reef Invertebrates and Plants, Patrick L. Colin. T. F. H. Publications, Neptune City, New Jersey. 1978.

A Field Guide to Coral Reefs of the Caribbean and Florida, Eugene H. Kaplan. Houghton Mifflin Company, Boston. 1982.

A Guide to the Natural History of the Cayman Islands, Nancy Sefton. Cayman Islands Conservation Association, Grand Cayman, BWI. 1976.

INDEX